T0233656

Renewable Energy Integration with Building Energy Systems

Construction, as an industry sector, is responsible for around one-third of the total world-wide energy usage and about 20% of greenhouse gas emissions. The rise in the number of buildings and floor space area for residential and commercial purposes has imposed enormous pressure on existing energy sources. Implementations such as efficient usage of building energy systems, design measures, utilization of local energy resources, energy storage, and the use of renewable energy sources to meet electricity demands are currently under development and deployment for improving the energy performance index. However, integrating all such measures and the development of nearly zero-energy and zero-emission buildings is yet to be explored.

In this book, the different control techniques and intelligent technologies used to improve the energy performance of buildings are illustrated. Every building energy control system has a two-fold objective for energy and comfort requirements to achieve a high comfort index (for thermal, visual, air quality, humidity, and various plug loads) and increase the energy performance index. The most significant aspect in the design of a building's energy control system is modelling. All the components, methodologies, and processes involved in developing a renewable energy-driven building are covered in detail.

This book is intended for graduates and professionals working towards the development of a sustainable built environment using renewable energy sources.

V.S.K.V. Harish is an Assistant Professor with the Department of Electrical Engineering at Netaji Subhas University of Technology (NSUT), Dwarka, New Delhi, India. He completed his ME (with Gold Medal) in Power Engineering from Jadavpur University in 2012 and his PhD from IIT Roorkee in 2017, both with an MHRD fellowship. After his PhD, he worked as a Post-Doctoral Fellow at the TERI School of Advanced Studies, New Delhi, under the Netherlands' government-sponsored project on smart grids for rural India. He was nominated for the prestigious Young Energy Researcher Award in 2016 and in 2018 at World Sustainable Energy Days, Austria. He was awarded the Best Paper Award at the Springer and IEEE International conferences in 2019. He was granted international travel support by the Department of Science and Technology (DST), Government of India, in 2014 and 2018, and national travel support by Soft Computing Research Society (SCRS), India, in 2017. He has published several research papers in journals and conferences of international repute. He has published book chapters with CRC Press/Taylor & Francis and Springer. He is also a member of various societies and institutions, such as ASHARE, IEEE, SCRS, CIBSE, etc. His research interests include building energy systems and power systems analysis.

Amit Vilas Sant is Assistant Professor with the Department of Electrical Engineering, School of Technology, Pandit Deendayal Petroleum University, Gandhinagar. He completed his PhD at the Indian Institute of Technology Delhi, New Delhi, in 2013. Prior to this, he completed his Master of Technology in Power Apparatus and Systems (with Gold Medal) from Nirma

University, Ahmedabad, in 2007. He obtained his Bachelor of Technology (Electrical and Electronics Engineering) and diploma in Electrical Engineering from Manipal Academy of Higher Education, Manipal, and The Maharaja Sayajirao University of Baroda, Vadodara. From April 2012 to September 2014, Dr. Amit was a post-doctoral researcher at the Masdar Institute of Science and Technology, Abu Dhabi. He has published several research papers in journals and conferences of international repute. Four of his research publications have received awards at different conferences. His research interests include power electronics, electric drives, electric vehicles, power quality enhancement and grid integration of renewable energy systems.

Arun Kumar has a bachelor's in civil engineering from IIT Roorkee, a Master's in Civil Engineering from IISc Bengaluru and a PhD from IIT Roorkee. He did Hydropower diploma studies at NTH, Trondheim, Norway. Prof. Kumar is currently working as a Professor at the Department of Hydro and Renewable Energy (HRED), IIT Roorkee. He held MNRE Chair Professor from 2013 to 2018, headed the department from 1998 to 2011, and served on the board of NHPC Ltd, Government of India PSU during 2015–2019 as independent director. His research areas are hydropower development, environmental management of water bodies, energy economics and policy. Prof. Arun Kumar regularly teaches various postgraduate/undergraduate courses at HRED, and students of other departments of the institute. He has designed two M.Tech programmes at HRED: Alternate Hydro Energy Systems and Environmental Management of Rivers and Lakes. He has also designed three optional courses for undergraduate students. He has contributed significantly to the establishment of the national and international recognition of the Department of Hydro and Renewable Energy. He has conceptualized and established national facilities for hydropower simulation, performance evaluation of hydro projects, hydraulic turbines, standards for SHP and environmental management of lakes and rivers, and comprehensive use of solar energy at the IIT Roorkee campus. Prof. Kumar has over 39 years of experience in the fields of hydropower and the environmental management of rivers and lakes. He served CLA for Hydropower on SRREN for IPCC. He appraises project proposals for the Government of India regularly. Prof. Kumar has received various awards and recognitions from CBIP, the Institution of Engineers, and the Hydropower Association.

Renewable Energy Integration with Building Energy Systems

A Modelling Approach

Edited by

V.S.K.V. Harish
Department of Electrical Engineering,
Netaji Subhas University of Technology (NSUT),
New Delhi, India

Amit Vilas Sant
Department of Electrical Engineering,
School of Technology,
Pandit Deendayal Petroleum University,
Gandhinagar, India

Arun Kumar
Department of Hydro and Renewable Energy,
Indian Institute of Technology Roorkee,
Roorkee, India

CRC Press
Taylor & Francis Group
Boca Raton London New York

CRC Press is an imprint of the
Taylor & Francis Group, an **informa** business

MATLAB® is a trademark of The MathWorks, Inc. and is used with permission. The MathWorks does not warrant the accuracy of the text or exercises in this book. This book's use or discussion of MATLAB® software or related products does not constitute endorsement or sponsorship by The MathWorks of a particular pedagogical approach or particular use of the MATLAB® software.

First published 2023
by CRC Press/Balkema
Schipholweg 107C, 2316 XC Leiden, The Netherlands
e-mail: enquiries@taylorandfrancis.com
www.routledge.com – www.taylorandfrancis.com

CRC Press/Balkema is an imprint of the Taylor & Francis Group, an informa business

Library of Congress Cataloging-in-Publication Data
A catalog record has been requested for this book

ISBN: 978-1-032-07488-7 (hbk)
ISBN: 978-1-032-07798-7 (pbk)
ISBN: 978-1-003-21158-7 (ebk)

DOI: 10.1201/9781003211587

Typeset in Times New Roman
by Newgen Publishing UK

Contents

Preface

Buildings, as a sector, are reported to be responsible for around 40% of total carbon-related emissions and over 33% of final energy usage, globally. Driven by improved standard of living, rising population and increased dependency on energy driven services, the energy demand from the buildings and construction sector is forecasted to rapidly grow in future. In order to meet this increased demand with reduced dependency on emission oriented conventional technologies; adoption of energy efficient and green energy systems are necessary. With a focus on sustainable development, there is an increasing interest in commercialization and wider implementation of electric vehicles and renewable energy technologies. Developing renewable energy systems integrated with building energy systems, possess high potential to support energy efficient techniques and energy conservation measures.

Integration of renewable energy with building energy systems facilitate improved energy system performance with enhanced efficiencies, utilization of on-site renewable energy resources due to availability of large surface areas within the built environment, decreased distribution system losses, and enhances usage of e-mobility such as electric vehicles. Primary objective of the book is to provide an extensive overview of the building energy systems and associative renewable energy technologies coupled with integration methodologies and power electronic components required for interfacing.

Improving the energy performance of a building involves development of energy monitoring and control strategies, which in turn, depend, significantly, on building energy models. Chapter 1 introduces significant terms for understanding the hygro-thermal fundamentals of the building energy physics. Fundamental mathematical formulation for basic heat and mass transfer processes are described in detail. Detailed discussion on various modes of thermal energy transfer such as conduction, convection and radiation, in building energy systems is also presented. The thermal electrical analogy which will be utilized throughout this book is also explained.

A building energy system comprises of the building construction elements, hygro-thermal energy transfer processes or a combination of the physical elements and processes. Chapter 2 deals with illustration of building energy system model development and techniques to analyze the developed energy models. Energy transfer equations for a multi-layered building construction element are represented and thermal resistor-capacitor networks are used to represent the element model. A single zonal energy system, integrating all the developed element models, as resistor-capacitor network is then developed. Step response for the developed un-conditioned building energy system model is analyzed in terms of the performance characteristics.

In Chapter 3, an energy model is developed for a Heating Ventilation and Air-Conditioning (HVAC) system. A simple, all-seasonal HVAC system is considered for the present study incorporating five-loops of operation viz., airside loop, chilled water loop, refrigeration loop, heat rejection loop and control loop. Mathematical formulations are developed for the primary components, such as air handling unit (consisting of mixing box, heating coil, humidifier, cooling coil and dehumidifier, fan and ductwork) and an air-conditioned single zonal room, of a HVAC system. MATLAB/Simulink has been used to develop the HVAC system model.

Since past few decades, high voltage dc transmission, flexible ac transmission, custom power devices and industrial drives have emerged and been incorporated in utility networks and industrial plants. Chapter 4 introduces the primary power electronic interface components required for renewable energy integration of building energy systems. Power converters, developed with the help of power semiconductor switches, serve to modulate the power for the desired load operation without compromising the efficiency. Power switches and power converters are becoming more and more compact and more and more efficient. The power converters, which can perform ac–dc, dc–ac, dc–dc and ac–ac power conversion, find applications in low power applications such as dc power supply to wind energy conversion systems rated in MW capacity.

Chapter 5 discusses the different aspects of grid integration of solar and wind energy. Also, the use of battery storage is also discussed. The related statistical data is presented as well. Wind and solar energy are available in abundance at different locations and is free of cost. In buildings, rooftop solar system can be installed. Similarly, vertical axis wind turbines can also be installed in the building premises. This can reduce the energy bills as well as carbon emissions from the building sector.

Chapter 6 showcases the significance of Electric Vehicles (EVs) in today's scenario. The electric propulsion system, its working and the components involved are discussed. Charging infrastructure is a major component that needs significant consideration for ensuring wider acceptance of the EV technology. This chapter covers the different methods of charging an electric vehicle battery. Different levels of charging are also included in the discussion along with the practical implementation and impact of electric vehicle battery changing on the grid operation.

Techno-economic analysis of electric vehicles is presented in Chapter 7 to demonstrate its merits and cost competitiveness over the period of operation. The statistics pertaining to the usage of EVs across the globe and India in particular are also stated. The merits and challenges for electric vehicles are discussed. A detailed cost comparison of two-wheeler electric vehicles against its internal combustion engine counterpart, in terms of initial and operating cost, is presented to show the cost competitiveness of the electric vehicles.

Contributors

Karan Doshi, Department of Electrical Engineering, School of Technology, Pandit Deendayal Energy University (PDEU), Raisan, Gandhinagar, Gujarat, India.

V.S.K.V. Harish, Department of Electrical Engineering, Netaji Subhas University of Technology (NSUT), New Delhi, India.

Arun Kumar, Department of Hydro and Renewable Energy, Indian Institute of Technology Roorkee, Roorkee Uttarakhand, India.

Chaitali Mehta, Department of Computer Science Engineering, School of Technology, Pandit Deendayal Energy University, Gandhinagar, India.

Arpit J. Patel, Department of Electrical Engineering, School of Technology, Pandit Deendayal Energy University (PDEU), Raisan, Gandhinagar, Gujarat, India.

Meet Patel, Department of Electrical Engineering, School of Technology, Pandit Deendayal Energy University (PDEU), Raisan, Gandhinagar, Gujarat, India.

Ojaswini A. Sharma, Department of Electrical Engineering, School of Technology, Pandit Deendayal Energy University (PDEU), Raisan, Gandhinagar, Gujarat, India.

Nayan Kumar Singh, Department of Electrical Engineering, School of Technology, Pandit Deendayal Energy University (PDEU), Raisan, Gandhinagar, Gujarat, India.

Amit Vilas Sant, Department of Electrical Engineering, School of Technology, Pandit Deendayal Energy University (PDEU), Raisan, Gandhinagar, Gujarat, India.

Chapter 1

Fundamentals of energy transfer in buildings

V.S.K.V. Harish, Nayan Kumar Singh, Arun Kumar,
Karan Doshi, and Amit Vilas Sant

CONTENTS

1.1 INTRODUCTION

Understanding energy transfer processes together with advanced technologies can help in the development of effective energy control strategies and improve the energy performance of buildings. It is imperative to define goal-specific energy control targets for the building energy modeller, which then form basic entities in the development of a building energy management system. Building information modelling (BIM) is an effective tool for analyzing energy behaviour with a geometric visual representation of the building under study. With the use of BIM software tools, a three-dimensional representation of the electro-mechanical and architectural design features of a building can be represented and analyzed easily. Moreover, BIM, integrated with smart metering and monitoring technologies, ensures the optimum (or near-optimum) operational performance of a building, giving due consideration to the occupants' comfort and the equipment's safety.

Before developing an energy system model, it is imperative to understand and gain knowledge of energy transfer processes, including hygro-thermal processes, in buildings. Knowledge of psychometrics, which involves mathematical formulae depicting the relationships of perfect gas and the thermodynamic properties of air, is also significant.

Some important terms for understanding the hygro-thermal fundamentals of building energy physics are illustrated in Table 1.1.

DOI: 10.1201/9781003211587-1

Table 1.1 Nomenclature for significant terms used in hygro-thermal dynamics modelling

Name	Symbol	Description	Units
Thermal parameters			
Heat	Q	The energy transferred across a system boundary by temperature difference.	Joule, J
Heat flow rate	$\dot{Q}$	The rate of transfer of heat.	J/s or watt (W)
Heat flux	φ	The heat flow rate through a surface.	W
Heat flux density	q	The heat flux per unit area.	W/m^2
Heat capacity	C	The amount of heat energy needed to raise the temperature of a substance by $1\,°C$. Also called thermal mass or thermal capacitance.	J/K
Specific heat capacity	Cp	The amount of heat energy needed to raise the temperature of 1 kg of material by $1\,°C$.	J/kg K
Mass flow rate	$\dot{m}$	The rate of change of mass per unit time.	kg/s
Volume flow rate	$\dot{V}$	The rate of change of volume per unit time.	m^2/s
Humidity parameters			
Humidity/mixing ratio	W	The ratio of water vapour mass to that of dry air for moist air: $$W = \frac{M_{wat}}{M_{dry}} \qquad (1.1)$$ $$= 0.621945 \frac{x_{wat}}{x_{dry}} \qquad (1.2)$$	
Specific humidity	γ	The ratio of the mass of water vapour to the total mass of the moist air sample: $$\gamma = M_w / (M_w + M_{da}) \qquad (1.3)$$ In terms of humidity ratio, $$\gamma = W / (1 + W) \qquad (1.4)$$	
Absolute humidity (water vapour density)	d_v	The ratio of the mass of the water vapour to the total volume of the sample: $$d_v = M_w / V \qquad (1.5)$$	
Density	ρ	The density of a moist air mixture is the ratio of the total mass to total volume: $$\rho = (M_{da} + M_w)/V = (1/v)(1 + W) \qquad (1.6)$$ where v is the moist air specific volume, m^3/kg_{da}, as defined by the equation.	
Saturation humidity ratio	$W_s(t,p)$	The humidity ratio of moist air saturated with respect to water (or ice) at the same temperature t and pressure p.	
Relative humidity	ϕ	The ratio of the actual water vapour partial pressure in moist air at the dew-point pressure and temperature to the reference saturation water vapour partial pressure at the dry-bulb pressure and temperature: $$\phi = (p_{wv-enh} / p_{wvs-ref} \mid_{p,t})$$ $$= \left[f(p,t_{dp})e(t_{dp}) \right] / \left[f(p,t_{db})e(t_{db}) \right] \qquad (1.7)$$	

Table 1.1 (continued)

Name	Symbol	Description	Units
Dew-point temperature	t_d	The temperature of the moist air saturation at pressure p, with the same humidity ratio W as that of the given sample of moist air. It is defined as the solution $t_d(p,W)$ of the following equation: $W_s(p,t_d) = W$ ⠀⠀⠀⠀⠀⠀⠀⠀⠀(1.8)	
Thermodynamic wet-bulb temperature	t^*	The temperature at which water (liquid or solid), by evaporating into moist air at dry-bulb temperature t and humidity ratio W, can bring air to saturation adiabatically at the same temperature t^* while total pressure p is constant.	

1.2 ENERGY TRANSFER IN BUILDING ENERGY SYSTEMS

The mathematical formula for basic heat transfer is given as:

$$h_{12} = m_{air}\left(h_2 - h_1\right) \tag{1.9}$$

$$h_{12} = m_{air}C_p\left(T_2 - T_1\right) \tag{1.10}$$

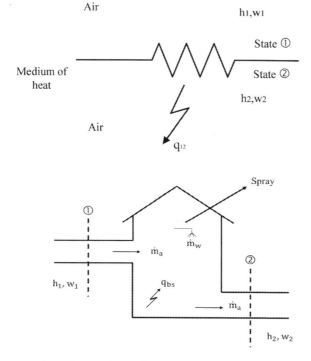

Figure 1.1 Mass flow rate through a conditioned building space.

The governing mathematical formulae can be deduced as:

$$\dot{m}_a h_1 + q_{bs} + q_w = \dot{m}_a h_2 \tag{1.11}$$

and

$$\dot{m}_a h_1 + q_{bs} + \Sigma\left(\dot{m}_w h_w\right) = \dot{m}_a h_2 \tag{1.12}$$

Also:

  $\forall$ Adiabatic mixing

$$\dot{m}_a h_1 + \dot{m}_w h_w = \dot{m}_a h_2 \tag{1.13}$$

Dividing by h_w throughout:

$$\dot{m}_a \frac{h_1}{h_w} + \dot{m}_w = \dot{m}_a \frac{h_2}{h_w} \tag{1.14}$$

$$\dot{m}_a H_1 + \dot{m}_w = \dot{m}_a H_2 \tag{1.15}$$

Therefore:

$$\frac{h_2 - h_1}{H_2 - H_1} = \frac{\Delta h}{\Delta H} = h_w \tag{1.16}$$

Also:

$$\dot{m}_a H_1 + \Sigma \dot{m}_w = \dot{m}_a H_2 \tag{1.17}$$

$$q_{bs} + h_w \Sigma \dot{m}_w = \dot{m}_a \left(h_2 - h_1\right) \tag{1.18}$$

or

$$\Sigma \dot{m}_w = \dot{m}_a \left(H_2 - H_1\right) \tag{1.19}$$

Here:

$$\frac{h_2 - h_1}{H_2 - H_1} = \frac{\Delta h}{\Delta H} = \frac{q_{bs} + h_w \Sigma \dot{m}_w}{\Sigma \dot{m}_w} \tag{1.20}$$

1.2.1 Modes of thermal energy transfer in building energy systems

Thermal energy in the form of heat is transferred from a high-temperature region to a lower temperature region. This energy transfer takes place in three modes: conduction, convection

Table 1.2 Laws of thermodynamics

Law	Statement	Mathematical depiction
Zeroth	If two bodies are in thermal equilibrium with a third body, they are also in thermal equilibrium with each other.	If $T_A = T_B$ and $T_B = T_C$, then $T_C = T_A$
First	Energy can neither be created nor destroyed, it can only be converted (or transformed) from one form to another, and the total energy remains constant.	$E = E_{in} - E_{out}$
Second	It is impossible for a heat engine to produce a network in a complete cycle if it exchanges heat only with bodies at a single fixed temperature. (Kelvin Planck) It is impossible to construct a device operating in a cycle that can transfer heat from a colder body to warmer without consuming any work. (Clausius)	$\Delta S = \int \dfrac{dq}{T}$

and radiation. Conduction is the transfer of energy from the more energetic particles of a substance to the adjacent less energetic ones as a result of the interaction between particles. Convection is the transfer of energy between a solid surface and the adjacent fluid that is in motion and involves the combined effects of conduction and fluid motion. Radiation is the transfer of energy due to the emission of electromagnetic waves (or photons). The laws of thermodynamics (Table 1.2) play a significant role in defining the fundamentals of building energy transfer principles.

Thermal energy transfer can be computed as:

$$Heat\ energy = \frac{mass \times specific\ heat}{Total\ thermal\ mass} \times (Temperature\ difference) \tag{1.21}$$

In a building energy system, a zone has a physical boundary that is the walls and surfaces of the zone, and the only thing that flows into and out of the zone (and interacts with the environment) is heat flow. Every room (/zone) is its own system, and the only input and outputs from the system that interact with the environment are heat flows, which is how the environment exerts its influence on the system. So the whole idea of modelling the building comes from defining two things: first, the boundaries of the zone, which is called the envelope of a zone or a building envelope; and second, the in and outflows of these boundaries. If we can mathematically describe these two things then we can design a good model of a zone and predict its temperature.

There are different types of heat gains, one of which is called sensible gain or sensible cooling. This refers to anything that can directly influence the zone air temperature because we can immediately sense it. Another is latent gain or latent cooling, which refers to anything that indirectly gives discomfort, such as a change in humidity.

In the room of a building, heat flows through all possible surfaces such as walls, windows, doors, roof and floor. The air inside the room also exchanges heat with every surface through a convection process. There are also internal heat gains that are generated through metabolic activities, and computers and lights generating heat. All of these accumulated processes are responsible for the generation of these energy or heat flows and occur through processes of different heat transfer based on heat through surfaces or walls, or heat flow through zone air or gas.

1.3 THERMAL ENERGY TRANSFER THROUGH A BUILDING CONSTRUCTION ELEMENT (WALL/SLAB)

Consider a wall, as shown in Figure 1.2, depicting, primarily, conduction heat transfer through an external wall (wall subjected to outdoor environment).

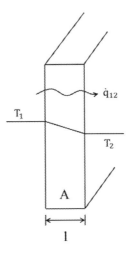

Figure 1.2 A building construction element (multi-layered wall/slab).

Consider:

$$T_1 > T_2$$

Then:

$$q_{12} = \frac{kA}{l}(T_1 - T_2) \tag{1.22}$$

$$= \frac{T_1 - T_2}{\left(\dfrac{l}{kA}\right)} \tag{1.23}$$

Conductive thermal resistance:

$$R_{cond} = \frac{l}{kA} \tag{1.24}$$

Figure 1.3 Conductive thermal resistance (R_{cond}).

In Figure 1.2, on the surfaces of each side of the wall, there exists convective heat exchange with the surrounding air and radiant heat exchanges with surfaces other than those which are exposed. For a roof or an external wall, the radiant heat exchange at the external side consists of absorbed solar radiation, which includes both diffuse and direct radiation. Theoretically, the convection mode of heat transfer can be represented as shown in Figure 1.4.

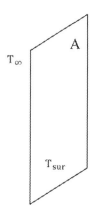

Figure 1.4 Convective heat transfer through a surface plane.

Convective heat is defined, mathematically, as:

$$q_{conv} = h_c A\left(T_{sur} - T_\infty\right) \tag{1.25}$$

$$= \frac{T_{sur} - T_\infty}{\left(\dfrac{1}{h_c A}\right)} \tag{1.26}$$

Convective thermal resistance:

$$R_{conv} = \frac{1}{h_c A} \tag{1.27}$$

$$\forall, T_\infty > T_{sur},$$

$$q_{conv} = h_c A \left(T_\infty - T_{sur} \right) \tag{1.28}$$

Theoretically, a radiative mode of heat transfer can be represented as shown in Figure 1.5.

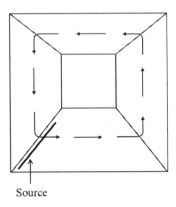

Source

Figure 1.5 Radiative heat transfer through a surface plane.

Combined radiation and convective heat are defined mathematically as:

$$q_{rad/cov} = q_{rad} + q_{conv} \tag{1.29}$$

$$= A_{sur} \left(h_{rad} + h_{conv} \right) \left(T_{sur} - T_\infty \right) \tag{1.30}$$

The heat transfer equation for a homogeneous material can be derived by considering the following assumptions:

- The rate of heat transfer across each boundary surface of an elemental control volume within the material that would arise corresponding to the temperature gradient that exists at the surface.
- Internal heat sources or sinks are present inside the control volume.
- The change in internal energy of the material in the control volume is reflected by the change in temperature of the material.

Applying the assumptions, we get the following equation:

$$\rho c \frac{\partial T}{\partial t} - \nabla \cdot k \nabla T - \dot{q} = 0 \tag{1.31}$$

To analyze heat transfer through walls and slabs in buildings, the above equation is further simplified by considering the following assumptions:

1. The heat transfer is only in one direction, that is, across the thickness of a wall or a slab, and it will be modelled in that one direction only. The heat transfer in the other two directions is neglected.
2. The heat transfer in the material is isotropic.

3. Material properties such as thermal conductivity, density and specific heat are independent of temperature.
4. No internal heat source or sink is present inside the material.

The governing equation can then be simplified to:

$$\frac{\partial T}{\partial t} = \alpha \frac{\partial^2 T}{\partial x^2} \qquad (1.32)$$

where $= k / \rho c$; $\alpha =$ thermal diffusivity of the material.
  $k =$ thermal conductivity of the material
  $\rho =$ density of the material
  $c =$ specific heat of the material

1.4 THERMAL ENERGY TRANSFER THROUGH A BUILDING CONSTRUCTION ELEMENT (WINDOW)

For transparent building construction elements such as windows, when incident solar radiation falls on the window glass, some part of the incident solar radiation is transmitted through the window glass pane inside the indoor space, and the remaining part is reflected from the surface of the window glass pane. Some of the energy from the transmitted solar radiation will be absorbed by the window glass, as a result of which the temperature of the glass increases. Due to this increase in glass temperature, the heat will flow in both the indoor and the outdoor directions. First, by conduction within the glass, and then by convection and radiation at both sides and surfaces of the glass pane.

Apart from the external walls and roofs, windows and skylights are the key components of a building envelope, through which heat transfer takes place between the indoor space and outdoor space. Window glass is generally much thinner and more transparent as compared to the external walls, and therefore, solar radiation penetration occurs easily in windows. In fact, it dominates the heat gain or loss of a heated or air-conditioned space in the building.

The heat transfer processes that take place in a window glass pane include:

1.1 Reflection, absorption and transmission of direct and diffused solar radiation.
1.2 Conduction and convection of the absorbed solar radiation with the ambient air.
1.3 Conduction and convection due to a temperature difference between indoor space and outdoor space.
1.4 Radiant heat exchange with the internal surfaces of other fabric elements and with lighting, appliances and other heat sources or sinks in the room.

Humans spend more than 90% of their time in buildings, and the buildings themselves are designed to provide a comfortable indoor environment for occupants. Human-building interactions, such as the usage of lighting and air conditioning, consume around one quarter to half of the total amount of commercial building energy. Office workers arrive and leave the workspace regularly according to schedules. Therefore, to reduce the energy consumption caused by human-building interaction while maintaining its occupants' comfort, it is necessary to utilize occupancy information. The occupancy schedules used to optimize building operations are conservative on energy savings. Larger savings with stable comfort satisfaction can be further achieved through learning and prediction of office occupancy.

1.5 THERMAL-ELECTRICAL ANALOGY

The study and analytical investigation of physical behaviour for mechanical, electrical, thermal and optical systems is time-consuming due to the complex mathematical formulae involved in the design problem. Analogues are used to represent complex systems and use simpler systems as a means of investigating the behaviour of the other system. Steady state and transient heat transfer problems are represented for modelling and simulation using electrical circuits due to their ease in assembly and instrumentation. Modelling using the electrical analogy involves mapping construction properties and climate data into electrical component values and interpreting electrical modelling results in terms of building properties, energy and climate. The relationship between thermal and electrical parameters for modelling a thermo-electrical circuit is represented in Table 1.3.

Table 1.3 Thermal-electrical analogy

Thermal			Electrical		
Parameter	*Symbol*	*S.I. unit*	*Parameter*	*Symbol*	*S.I. unit*
Temperature	T, θ	°C, K	Voltage	V	Volt
Time	t	S	Time	t	S
Heat-flow rate	$\dfrac{\partial q}{\partial t}$	W	Current	i	A
Heat capacity	ρ	J/(kg-°C), J/(kg-K)	Capacitance	C	°C/m³
Conductivity	k	W/(m-K), W/(m-°C)	Conductivity	$1/R$	Mho
Length	l, x	m	Length	l, x	M
Temperature gradient	$\dfrac{\partial T}{\partial x}, \dfrac{\partial \theta}{\partial x}$	°C/m, K/m	Voltage gradient	$\dfrac{\partial V}{\partial x}$	Volt/m
Rate of temperature rise	$\dfrac{\partial T}{\partial t}, \dfrac{\partial \theta}{\partial t}$	°C/s, K/s	Rate of voltage rise	$\dfrac{\partial V}{\partial t}$	Volt/s

A typical 3R2C model for a building construction element is shown in Figure 1.6.

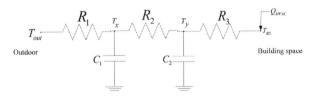

Figure 1.6 RC-network model describing thermal energy transfer.

Ordinary differential equations (ODEs) for a 3R2C network of a second-order multi-layered building element is given as equation (1.33) to equation (1.36).

$$C_1 \frac{d(T_x)}{dt} = \left(\frac{(T_{out} - T_x)}{R_1} - \frac{(T_x - T_y)}{R_2} \right) \tag{1.33}$$

$$\Rightarrow \frac{d(T_x)}{dt} = \frac{1}{C_1} \left(\frac{(T_{out} - T_x)}{R_1} - \frac{(T_x - T_y)}{R_2} \right) \tag{1.34}$$

and

$$C_2 \frac{d(T_y)}{dt} = \left(\frac{(T_x - T_y)}{R_2} - \frac{(T_y - T_{BS})}{R_3} + Q_{HVAC} \right) \tag{1.35}$$

$$\Rightarrow \frac{d(T_y)}{dt} = \frac{1}{C_2} \left(\frac{(T_x - T_y)}{R_2} - \frac{(T_y - T_{BS})}{R_3} + Q_{HVAC} \right) \tag{1.36}$$

where

T_x	=	Nodal temperature representing the effective outer surface of multi-layered building element under study, $°C$
T_y	=	Nodal temperature representing the effective inner surface of multi-layered building element under study, $°C$
T_{out}	=	Outdoor or external air temperature, $°C$
T_{BS}	=	Building space air temperature, $°C$
Q_{HVAC}	=	HVAC plant heat rate, W
R_1	=	$x_1 \times R_{th}$
R_2	=	$x_2 \times R_{th}$
R_3	=	$x_3 \times R_{th}$
C_1	=	$y_1 \times C_{th}$
C_2	=	$y_2 \times C_{th}$
R_{th}	=	Total thermal resistance of multi-layered building element, $m^2\,°C/W$
C_{th}	=	Total thermal capacitance of multi-layered building element, $J/m_2\,°C$

Equation (1.34) and equation (1.36) are re-written in matrix form as equation (1.37).

$$\begin{pmatrix} \dot{T}_x \\ \dot{T}_y \end{pmatrix} = \begin{pmatrix} -\left\{ \frac{1}{x_1 R_{th}} + \frac{1}{x_2 R_{th}} \right\} \frac{1}{y_1 C_{th}} & \frac{1}{x_2 y_1 R_{th} C_{th}} \\ \frac{1}{x_2 y_2 R_{th} C_{th}} & -\left\{ \frac{1}{x_2 R_{th}} + \frac{1}{x_3 R_{th}} \right\} \frac{1}{y_2 C_{th}} \end{pmatrix} \begin{pmatrix} T_x \\ T_y \end{pmatrix} + \dots$$

$$\begin{pmatrix} \frac{1}{x_1 y_1 R_{th} C_{th}} & 0 & 0 \\ 0 & \frac{1}{y_2 C_{th}} & \frac{1}{x_3 y_2 R_{th} C_{th}} \end{pmatrix} \begin{pmatrix} T_{out} \\ Q_{hvac} \\ T_{BS} \end{pmatrix} \tag{1.37}$$

1.6 NUMERICAL EXAMPLES

Example   Consider a single zonal room of 10m × 10m made up of a building element of 10″ thickness and thermal conductivity of 0.75 w/m-K. The outdoor and indoor surface temperatures of the room are maintained at 30°C and 20°C, respectively. Determine the rate of heat transfer from the room through the construction element.

Solution

Geometrically representing the room as a cuboid (10m ×10m, 6m height):

Area:

$$A_{sur} = 2(10 \times 10 + 10 \times 6 + 10 \times 6)$$
$$= 440 \, \text{m}^2$$

$$q_{wall} = \frac{kA_{sur}}{l}(T_{out} - T_{in})$$
$$= \frac{0.75 \times 440}{0.254} \times 10$$
$$= 16.5 \, \text{KW}$$

Example   Consider a single-layered construction element made of bricks. The thickness of the 5m × 6m element is 30cm. If the outdoor and indoor surface temperatures of the element are maintained at 30°C and 20°C, respectively, determine the rate of thermal energy transfer through the construction element (thermal conductivity, k = 0.79 w/m°C).

Solution

$$y = 6\text{m}; z = 5\text{m}; x = 30\text{cm}$$
$$T_{out} = 30°C; T_{in} = 20°C; k = 0.79 \text{w/m°C}$$

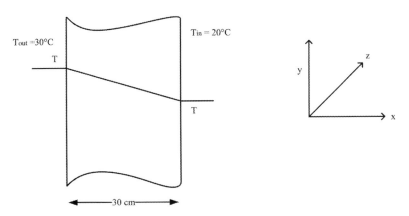

Figure 1.7 Temperature profile for single layered construction element.

Assuming heat rate along the x-axis (thickness):

$$q_x = -kA\frac{dT}{dx}$$

$$q_x = -kA\left(\frac{T_{out} - T_{in}}{30 - 0}\right)$$

Cross-sectional area:

$$A = x \times y = 5 \times 6 = 30 \text{ m}^2$$

Temperature gradient:

$$\Delta T = T_{out} - T_{in}$$
$$= 30°C - 20°C$$
$$= 10°C$$

Therefore:

$$q_x = -kA\frac{\Delta T}{\Delta x}$$

$$= -0.79 \times 30 \times \frac{10}{\frac{30}{100}}$$

$$= 790\,\text{W} \approx 790\,\text{J/s}$$

REFERENCES

ASHRAE (2017). *ASHRAE Handbook of Fundamentals*. American Society of Heating, Refrigerating, and Air-Conditioning Engineers. Atlanta, USA.

Cengel, Y. A., Boles, M. A., & Kanoglu, M. (2011). *Thermodynamics: An engineering approach* (Vol. 5, p. 445). New York: McGraw-Hill.

Harish, V.S.K.V., & Kumar, A. (2016). A review on modeling and simulation of building energy systems. *Renewable and Sustainable Energy Reviews*, 56, 1272–1292, doi.org/10.1016/j.rser.2015.12.040.

MASc, C. S., Lepine, C., & John Straube PhD, P. (2019). Trends and Anomalies in Hygrothermal Material Properties from the ASHRAE 1696 Research Program. In ASHRAE Topical Conference Proceedings (pp. 154–168). American Society of Heating, Refrigeration and Air Conditioning Engineers, Inc.

Sarfraz, O., & Bach, C. K. (2018). Equipment power consumption and load factor profiles for buildings' energy simulation (ASHRAE 1742-RP). *Science and Technology for the Built Environment*, 24(10), 1054–1063.

Chapter 2

Modelling and simulation of building energy elements

Nayan Kumar Singh, V.S.K.V. Harish, Arun Kumar, Karan Doshi, and Amit Vilas Sant

CONTENTS

2.1 INTRODUCTION

In a building, the system responsible for the energy consumption is the building energy system. This system can be anything such as machinery or any physical system, or a process or combination of them. The development of a building energy system model enables a better understanding of the energy transfer processes occurring within the building under study. This, in turn, helps the building designers/energy managers/analysts/auditors make decisions in a much more informed way, thereby enhancing effectiveness.

The output of a building energy system model can be categorized as: HVAC, lighting and other load designs; and energy performance and cost analysis. Load designing includes building space heating and cooling loads, volumetric ventilation needs, and equipment capacities, and so on. A building space's air-conditioning load is calculated by estimating the amount of thermal (heating/cooling) energy required for a particular building space under study. Ventilation needs are evaluated based on the number of air changes per hour and volumetric air flow requirements of a particular building space under study.

Energy performance and cost analysis is performed by estimating the energy usage within a simulation time period and the associative costs incurred. Based on the performance analysis, several energy efficient options can be compared, and avoidable carbon emissions can be calculated. Cost analysis also helps in performing a techno-economic analysis of the building energy system under study. Based on the load design and performance analysis, control strategies can be developed to improve the buildings' energy performance, maximizing the occupants' comfort in terms of indoor air-quality, visual, acoustic, and thermal comfort, and also safety. This has gained crucial importance in post-pandemic times as building HVAC designers and passive building developers look for strategies that can reduce the chances of virus transmission within a conditioned and unconditioned building space.

The most significant aspect of enhancing the performance of a building is developing a good building energy model. The modeller may decide on the simulation software package to be used depending upon the nature of the output of the building energy model. Two or more different softwares can be adopted depending upon the scale of the modelling project. A developed building energy model's performance should then be analyzed for stability and controllability before developing and testing the energy control strategies. This will help avoid the undesirable circumstance of attaining instability due to variations in any modelling parameter's behavior.

2.2 BUILDING SPACE/ZONE SYSTEM DESCRIPTION

Consider a building space consisting of the envelope of a room on the middle floor of a multistorey building. The space comprises an external wall, internal partition walls, window, ceiling, and floor. The air of the building space is conditioned using an HVAC system, which provides cooling or heating to the room by circulating air between the air handling unit and the room. The heat and mass transfer processes that take place in such a building space include:

- Conduction heat transfer through the room envelope, including the external walls, internal partition walls, window, ceiling, and floor.
- Solar radiation heat gain through window.
- Movement of the outdoor air and air from nearby rooms.
- Dissipation of heat and moisture inside the room, from the occupants, equipment, lighting, etc.
- Cooling or heating and dehumidification or humidification from the HVAC system.

A single zonal building energy system model is developed by considering the following assumptions for the energy modelling of multilayered construction elements:

1. Thermal energy balance is considered along the direction in which the thickness of the construction element is infinitesimally large, i.e., along the x direction. Thermal energy flow along y- and z-directions are negligible, and thus ignored.
2. All the imaginary internal planes parallel to the surfaces of the construction element are isothermal. This approximation neglects surface temperature gradients and edge effects.
3. Thermal energy transfer within the material is isotropic in nature.
4. The thermo-physical properties of the material such as density, specific heat, and thermal conductivity are constants and independent of temperature changes.
5. There is only the transfer of thermal energy, and no source or sink is present within the construction element.

The benefits of one-dimensional heat transfer offsets the inaccuracies introduced into the results by the isothermal approximation. Applying such assumptions, i.e., $\Delta x \rightarrow 0$ and $\Delta t \rightarrow 0$, the energy transfer equation is represented as equation (2.1) (see equation 1.32, Chapter 1).

$$\rho C \frac{\partial T}{\partial t} = k \frac{\partial^2 T}{\partial x_2} \tag{2.1}$$

$$=> \frac{\partial T}{\partial t} = \alpha \frac{\partial^2 T}{\partial x_2} \tag{2.2}$$

where

α = Thermal diffusivity of the material, (m^2/s)

$$\alpha = \frac{k}{\rho C}$$

k = Thermal conductivity of the material of building element, ($W/m\text{-}°C$)

A = Cross sectional area of building element, (m^2)

T_{BCE} = Temperature of the building element, ($°C$)

X = Thickness of building element, (m).

Integrating and rearranging equation (2.2), the heat transfer rate through a multilayered building element, ($\dot{Q}_{cond,BCE}$), is given as equation (2.3).

$$\dot{Q}_{cond,BCE} = -kA_{BCE} \frac{T_j - T_{j-1}}{L} \tag{2.3}$$

where

A_{BCE} = Cross-sectional area of the building element, (m^2)

T_j = Temperature at node, j, ($°C$)

T_{j-1} = Temperature at node, $j\text{-}1$, ($°C$)

L = Thickness of the building element, (m)

2.3 BUILDING CONSTRUCTION ELEMENT MODEL

A typical 3R2C model for a building construction element was discussed in Chapter 1. Now, a detailed representation of a construction element for a building space under study will be developed as a 3R2C (RC-network) model (Figure 1.6). Primarily, a building construction element is a composite of many layers. The thermal resistance of the construction element can be bifurcated, conceptually, by dividing the layers of the construction element into three parts. Each part will then possess an associative thermal resistance value with each resistor being a fraction of the total thermal resistance of that particular construction element. The thermal energy storage capacity of a multilayered construction element is represented as a thermal capacitance. A number of thermal capacitors can be employed to represent a building construction element, each capacitor being a fraction of the total thermal capacitance of that particular construction element.

Figure 2.1 shows the 3R2C thermal network model of a room in which there are two external walls, two internal walls, a glass window, a ceiling, and a floor. These are represented with there corresponding thermal resistance and thermal capacitance. T_{out}, Q_{ihg}, Q_{HVAC} and T_{BS} are the inputs of the model.

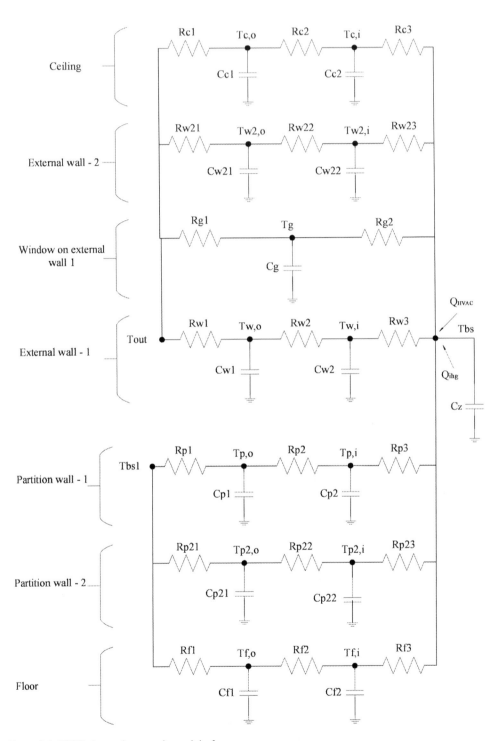

Figure 2.1 3R2C thermal network model of a room.

$T_{w,o}$ and $T_{w,i}$ are the outer surface and inner surface nodal temperature of external wall 1 in °C. By using these two nodal points, one can derive the ordinary differential equation for the 3R2C network of external wall 1. The energy balance equations for each node are developed as follows.

Energy balance equation for external wall 1
At node $T_{w,o}$

$$\frac{dT_{w,o}}{dt} = \frac{A_{w1}}{C_{w1}}\left(\left(\frac{T_{out} - T_{w,o}}{R_{w1}}\right) - \left(\frac{T_{w,o} - T_{w,i}}{R_{w2}}\right)\right) \tag{2.4}$$

$$\frac{dT_{w,o}}{dt} = \left(-\left(\frac{1}{R_{w1}} + \frac{1}{R_{w2}}\right)\frac{A_{w1}}{C_{w1}}\right)T_{w,o} + \left(\frac{A_{w1}}{C_{w1}R_{w2}}\right)T_{w,i} + \left(\frac{A_{w1}}{R_{w1}C_{w1}}\right)T_{out} \tag{2.5}$$

At node $T_{w,i}$:

$$\frac{dT_{w,i}}{dt} = \frac{A_{w1}}{C_{w2}}\left(\left(\frac{T_{w,o} - T_{w,i}}{R_{w2}}\right) - \left(\frac{T_{w,i} - T_{bs}}{R_{w3}}\right) + Q_{ihg} + Q_{HVAC}\right) \tag{2.6}$$

$$\frac{dT_{w,i}}{dt} = \left(\frac{A_{w1}}{R_{w2}C_{w2}}\right)T_{w,o} + \left(-\left(\frac{1}{R_{w2}} + \frac{1}{R_{w3}}\right)\frac{A_{w1}}{C_{w2}}\right)T_{w,i} + \left(\frac{A_{w1}}{R_{w3}C_{w2}}\right)T_{bs} + \left(\frac{A_{w1}}{C_{w2}}\right)$$
$$Q_{ihg} + \left(\frac{A_{w1}}{C_{w2}}\right)Q_{HVAC} \tag{2.7}$$

where

$T_{w,o}$	=	Outer surface nodal temperature of external wall 1, °C
$T_{w,i}$	=	Inner surface nodal temperature of external wall 1, °C
T_{out}	=	Outdoor air temperature, °C
T_{bs}	=	Building space air temperature, °C
Q_{HVAC}	=	HVAC plant heat rate, W
Q_{ihg}	=	Internal heat gain from occupant, light, and equipment in the building, W
A_{w1}	=	Area of external wall - 1, m²
R_{w1}	=	$x_1 \times R_{th,w1}$
R_{w2}	=	$x_2 \times R_{th,w1}$
R_{w3}	=	$x_3 \times R_{th,w1}$
C_{w1}	=	$y_1 \times C_{th,w1}$
C_{w2}	=	$y_2 \times C_{th,w1}$
$R_{th,w1}$	=	Total thermal resistance of external wall - 1, m² °C/W
$C_{th,w1}$	=	Total thermal capacitance of external wall - 1, J/m² °C

The energy balance equation for external wall 1 is shown in equation (2.4) and equation (2.5).

Energy balance equation for external wall 2

At node $T_{w2,o}$:

$$\frac{dT_{w2,o}}{dt} = \frac{A_{w2}}{C_{w21}}\left(\left(\frac{T_{out}-T_{w2,o}}{R_{w21}}\right)-\left(\frac{T_{w2,o}-T_{w2,i}}{R_{w22}}\right)\right) \tag{2.8}$$

$$\frac{dT_{w2,o}}{dt} = \left(-\left(\frac{1}{R_{w21}}+\frac{1}{R_{w22}}\right)\frac{A_{w2}}{C_{w21}}\right)T_{w2,o} + \left(\frac{A_{w2}}{C_{w21}R_{w22}}\right)T_{w2,i} + \left(\frac{A_{w2}}{R_{w21}C_{w21}}\right)T_{out} \tag{2.9}$$

At node $T_{w2,i}$

$$\frac{dT_{w2,i}}{dt} = \frac{A_{w2}}{C_{w22}}\left(\left(\frac{T_{w2,o}-T_{w2,i}}{R_{w22}}\right)-\left(\frac{T_{w2,i}-T_{bs}}{R_{w23}}\right)+Q_{ihg}+Q_{HVAC}\right) \tag{2.10}$$

$$\frac{dT_{w2,i}}{dt} = \left(\frac{A_{w2}}{R_{w22}C_{w22}}\right)T_{w2,o} + \left(-\left(\frac{1}{R_{w22}}+\frac{1}{R_{w23}}\right)\frac{A_{w2}}{C_{w22}}\right)T_{w2,i} + \left(\frac{A_{w2}}{R_{w23}C_{w22}}\right)$$
$$T_{bs} + \left(\frac{A_{w2}}{C_{w22}}\right)Q_{ihg} + \left(\frac{A_{w2}}{C_{w22}}\right)Q_{HVAC} \tag{2.11}$$

where

$$
\begin{array}{lll}
T_{w2,o} & = & \text{Outer surface nodal temperature of external wall 2, } ^\circ C \\
T_{w2,i} & = & \text{Inner surface nodal temperature of external wall 2, } ^\circ C \\
A_{w2} & = & \text{Area of external wall - 2, m}^2 \\
R_{w21} & = & x_1 \times R_{th,w2} \\
R_{w22} & = & x_2 \times R_{th,w2} \\
R_{w23} & = & x_3 \times R_{th,w2} \\
C_{w21} & = & y_1 \times C_{th,w2} \\
C_{w22} & = & y_2 \times C_{th,w2} \\
R_{th,w2} & = & \text{Total thermal resistance of external wall - 2, m}^2\,^\circ C/W \\
C_{th,w2} & = & \text{Total thermal capacitance of external wall - 2, J/m}^2\,^\circ C
\end{array}
$$

Energy balance equation for glass window

At node T_g:

$$\frac{dT_g}{dt} = \frac{A_g}{C_g}\left(\left(\frac{T_{out}-T_g}{R_{g1}}\right)-\left(\frac{T_g-T_{bs}}{R_{g2}}\right)+Q_{ihg}+Q_{HVAC}\right) \tag{2.12}$$

$$\frac{dT_g}{dt} = \left(-\left(\frac{1}{R_{g1}}+\frac{1}{R_{g2}}\right)\frac{A_g}{C_g}\right)T_g + \left(\frac{A_g}{R_{g1}C_g}\right)T_{out} + \left(\frac{A_g}{R_{g2}C_g}\right)T_{bs} + \left(\frac{A_g}{C_g}\right)$$
$$Q_{ihg} + \left(\frac{A_g}{C_g}\right)Q_{HVAC} \tag{2.13}$$

where

T_g	=	Surface nodal temperature of window, $°C$
A_g	=	Area of glass window, m^2
R_{g1}	=	$x_1 \times R_{th,g}$
R_{g2}	=	$x_2 \times R_{th,g}$
C_g	=	$C_{th,g}$
$R_{th,g}$	=	Total thermal resistance of glass window, $m^2 \, °C/W$,
$C_{th,g}$	=	Total thermal capacitance of glass window, $J/m^2 \, °C$.

Energy balance equation for ceiling

At node $T_{c,o}$:

$$\frac{dT_{c,o}}{dt} = \frac{A_c}{C_{c1}}\left(\left(\frac{T_{out}-T_{c,o}}{R_{c1}}\right)-\left(\frac{T_{c,o}-T_{c,i}}{R_{c2}}\right)\right) \tag{2.14}$$

$$\frac{dT_{c,o}}{dt} = \left(-\left(\frac{1}{R_{c1}}+\frac{1}{R_{c2}}\right)\frac{A_c}{C_{c1}}\right)T_{c,o}+\left(\frac{A_c}{C_{c1}R_{c2}}\right)T_{c,i}+\left(\frac{A_c}{R_{c1}C_{c1}}\right)T_{out} \tag{2.15}$$

At node $T_{c,i}$:

$$\frac{dT_{c,i}}{dt} = \frac{A_c}{C_{c2}}\left(\left(\frac{T_{c,o}-T_{c,i}}{R_{c2}}\right)-\left(\frac{T_{c,i}-T_{bs}}{R_{c3}}\right)+Q_{ihg}+Q_{HVAC}\right) \tag{2.16}$$

$$\frac{dT_{c,i}}{dt} = \left(\frac{A_c}{R_{c2}C_{c2}}\right)T_{c,o}+\left(-\left(\frac{1}{R_{c2}}+\frac{1}{R_{c3}}\right)\frac{A_c}{C_{c2}}\right)T_{c,i}+\left(\frac{A_c}{R_{c3}C_{c2}}\right)T_{bs}+\left(\frac{A_c}{C_{c2}}\right)$$
$$Q_{ihg}+\left(\frac{A_c}{C_{c2}}\right)Q_{HVAC} \tag{2.17}$$

where

$T_{c,o}$	=	Outer surface nodal temperature of ceiling, $°C$
$T_{c,i}$	=	Inner surface nodal temperature of ceiling, $°C$
A_c	=	Area of ceiling, m^2
R_{c1}	=	$x_1 \times R_{th,c}$
R_{c2}	=	$x_2 \times R_{th,c}$
R_{c3}	=	$x_3 \times R_{th,c}$
C_{c1}	=	$y_1 \times C_{th,c}$
C_{c2}	=	$y_2 \times C_{th,c}$

$R_{th,c}$ =   Total thermal resistance of ceiling, m² °C/W

$C_{th,c}$ =   Total thermal capacitance of ceiling, J/m² °C

Energy balance equation for partition wall 1

At node $T_{P,o}$:

$$\frac{dT_{P,o}}{dt} = \frac{A_P}{C_{P1}}\left(\left(\frac{T_{bs1} - T_{P,o}}{R_{P1}}\right) - \left(\frac{T_{P,o} - T_{P,i}}{R_{P2}}\right)\right)$$

$$\frac{dT_{P,o}}{dt} = \left(-\left(\frac{1}{R_{P1}} + \frac{1}{R_{P2}}\right)\frac{A_P}{C_{P1}}\right)T_{P,o} + \left(\frac{A_P}{C_{P1}R_{P2}}\right)T_{P,i} + \left(\frac{A_P}{R_{P1}C_{P1}}\right)T_{bs1} \qquad (2.18)$$

At node $T_{P,i}$:

$$\frac{dT_{P,i}}{dt} = \frac{A_P}{C_{P2}}\left(\left(\frac{T_{P,o} - T_{P,i}}{R_{P2}}\right) - \left(\frac{T_{P,i} - T_{bs}}{R_{P3}}\right) + Q_{ihg} + Q_{HVAC}\right)$$

$$\frac{dT_{P,i}}{dt} = \left(\frac{A_P}{R_{P2}C_{P2}}\right)T_{P,o} + \left(-\left(\frac{1}{R_{P2}} + \frac{1}{R_{P3}}\right)\frac{A_P}{C_{P2}}\right)T_{P,i} + \left(\frac{A_P}{R_{P3}C_{P2}}\right)T_{bs} + \left(\frac{A_P}{C_{P2}}\right)$$

$$Q_{ihg} + \left(\frac{A_P}{C_{P2}}\right)Q_{HVAC} \qquad (2.19)$$

where

$T_{p,o}$ =   Outer surface nodal temperature of partition wall 1, °C

$T_{p,i}$ =   Inner surface nodal temperature of partition wall 1, °C

T_{bs1} =   Building space air temperature 1, °C

A_p =   Area of partition wall - 1, m²

R_{p1} = $x_1 \times R_{th,p1}$

R_{p2} = $x_2 \times R_{th,p1}$

R_{p3} = $x_3 \times R_{th,p1}$

C_{p1} = $y_1 \times C_{th,p1}$

C_{p2} = $y_2 \times C_{th,p1}$

$R_{th,p1}$ =   Total thermal resistance of partition wall - 1, m² °C/W

$C_{th,p1}$ =   Total thermal capacitance of partition wall - 1, J/m² °C

Energy balance equation for partition wall 2

At node $T_{P2,o}$:

$$\frac{dT_{P2,o}}{dt} = \frac{A_{P2}}{C_{P21}}\left(\left(\frac{T_{bs1} - T_{P2,o}}{R_{P21}}\right) - \left(\frac{T_{P2,o} - T_{P2,i}}{R_{P22}}\right)\right)$$

$$\frac{dT_{P2,o}}{dt} = \left(-\left(\frac{1}{R_{P21}}+\frac{1}{R_{P22}}\right)\frac{A_{P2}}{C_{P21}}\right)T_{P2,o} + \left(\frac{A_{P2}}{C_{P21}R_{P22}}\right)T_{P2,i} + \left(\frac{A_{P2}}{R_{P21}C_{P21}}\right)T_{bs1} \qquad (2.20)$$

At node $T_{P2,i}$:

$$\frac{dT_{P2,i}}{dt} = \frac{A_{P2}}{C_{P22}}\left(\left(\frac{T_{P2,o}-T_{P2,i}}{R_{P22}}\right) - \left(\frac{T_{P2,i}-T_{bs}}{R_{P23}}\right) + Q_{ihg} + Q_{HVAC}\right)$$

$$\frac{dT_{P2,i}}{dt} = \left(\frac{A_{P2}}{R_{P22}C_{P22}}\right)T_{P2,o} + \left(-\left(\frac{1}{R_{P22}}+\frac{1}{R_{P23}}\right)\frac{A_{P2}}{C_{P22}}\right)T_{P2,i} + \left(\frac{A_{P2}}{R_{P23}C_{P22}}\right)T_{bs}$$

$$+ \left(\frac{A_{P2}}{C_{P22}}\right)Q_{ihg} + \left(\frac{A_{P2}}{C_{P22}}\right)Q_{HVAC} \qquad (2.21)$$

where

$T_{p2,o}$	=	Outer surface nodal temperature of partition wall 2, $°C$
$T_{p2,i}$	=	Inner surface nodal temperature of partition wall 2, $°C$
T_{bs1}	=	Building space air temperature 1, $°C$
A_{p2}	=	Area of partition wall - 2, m^2
R_{p21}	=	$x_1 \times R_{th,p2}$
R_{p22}	=	$x_2 \times R_{th,p2}$
R_{p23}	=	$x_3 \times R_{th,p2}$
C_{p21}	=	$y_1 \times C_{th,p2}$
C_{p22}	=	$y_2 \times C_{th,p2}$
$R_{th,p2}$	=	Total thermal resistance of partition wall - 2, $m^2\ °C/W$
$C_{th,p2}$	=	Total thermal capacitance of partition wall - 2, $J/m^2\ °C$

Energy balance equation for floor

At node $T_{F,o}$:

$$\frac{dT_{F,o}}{dt} = \frac{A_F}{C_{F1}}\left(\left(\frac{T_{bs1}-T_{F,o}}{R_{F1}}\right) - \left(\frac{T_{F,o}-T_{F,i}}{R_{F2}}\right)\right)$$

$$\frac{dT_{F,o}}{dt} = \left(-\left(\frac{1}{R_{F1}}+\frac{1}{R_{F2}}\right)\frac{A_F}{C_{F1}}\right)T_{F,o} + \left(\frac{A_F}{C_{F1}R_{F2}}\right)T_{F,i} + \left(\frac{A_F}{R_{F1}C_{F1}}\right)T_{bs1} \qquad (2.22)$$

At node $T_{F,i}$:

$$\frac{dT_{F,i}}{dt} = \frac{A_F}{C_{F2}}\left(\left(\frac{T_{F,o}-T_{F,i}}{R_{F2}}\right) - \left(\frac{T_{F,i}-T_{bs}}{R_{F3}}\right) + Q_{ihg} + Q_{HVAC}\right)$$

$$\frac{dT_{F,i}}{dt} = \left(\frac{A_F}{R_{F2}C_{F2}}\right)T_{F,o} + \left(-\left(\frac{1}{R_{F2}}+\frac{1}{R_{F3}}\right)\frac{A_F}{C_{F2}}\right)T_{F,i} + \left(\frac{A_F}{R_{F3}C_{F2}}\right)T_{bs} + \left(\frac{A_F}{C_{F2}}\right)$$
$$Q_{ihg} + \left(\frac{A_F}{C_{F2}}\right)Q_{HVAC} \tag{2.23}$$

where

$T_{F,o}$	$=$	Outer surface nodal temperature of floor, $^\circ C$
$T_{F,i}$	$=$	Inner surface nodal temperature of floor, $^\circ C$
T_{bsl}	$=$	Building space air temperature 1, $^\circ C$
A_F	$=$	Area of floor, m^2
R_{F1}	$=$	$x_1 \times R_{th,F}$
R_{F2}	$=$	$x_2 \times R_{th,F}$
R_{F3}	$=$	$x_3 \times R_{th,F}$
C_{F1}	$=$	$y_1 \times C_{th,F}$
C_{F2}	$=$	$y_2 \times C_{th,F}$
$R_{th,F}$	$=$	Total thermal resistance of partition wall - 2, $m^2\,^\circ C/W$
$C_{th,F}$	$=$	Total thermal capacitance of partition wall - 2, $J/m^2\,^\circ C$

Energy balance equation for building space air

$$\rho_a V c_a C_z \frac{dT_{bs}}{dt} = \left[A_c\left(\frac{T_{c,i}-T_{bs}}{R_{c3}}\right) + A_{w2}\left(\frac{T_{w2,i}-T_{bs}}{R_{w23}}\right) + A_g\left(\frac{T_g-T_{bs}}{R_{g2}}\right) + A_{w1}\left(\frac{T_{w,i}-T_{bs}}{R_{w3}}\right) \right.$$
$$\left. + A_P\left(\frac{T_{P,i}-T_{bs}}{R_{P3}}\right) + A_{P2}\left(\frac{T_{P2,i}-T_{bs}}{R_{P23}}\right) + A_F\left(\frac{T_{F,i}-T_{bs}}{R_{F3}}\right) + Q_{ihg} + Q_{HVAC} \right]$$

$$\frac{dT_{bs}}{dt} = \left(\frac{A_{w1}}{R_{w3}\rho_a V c_a C_z}\right)T_{w,i} + \left(\frac{A_{w2}}{R_{w23}\rho_a V c_a C_z}\right)T_{w2,i} + \left(\frac{A_g}{R_{g2}\rho_a V c_a C_z}\right)T_g$$
$$+ \left(\frac{A_c}{R_{c3}\rho_a V c_a C_z}\right)T_{c,i} + \left(\frac{A_p}{R_{P3}\rho_a V c_a C_z}\right)T_{P,i} + \left(\frac{A_{P2}}{R_{P23}\rho_a V c_a C_z}\right)T_{P2,i} +$$
$$+ \left(\frac{A_F}{R_{F3}\rho_a V c_a C_z}\right)T_{F,i} - \left(\frac{A_{w1}}{R_{w3}}+\frac{A_{w2}}{R_{w23}}+\frac{A_g}{R_{g2}}+\frac{A_c}{R_{c3}}+\frac{A_p}{R_{P3}}+\frac{A_{P2}}{R_{P23}}+\frac{A_F}{R_{F3}}\right)$$
$$\frac{T_{bs}}{\rho_a V c_a C_z} + \frac{Q_{ihg}}{\rho_a V c_a C_z} + \frac{Q_{HVAC}}{\rho_a V c_a C_z} \tag{2.24}$$

where

ρ_a	$=$	Density of indoor air, $kg\,m^{-3}$
V	$=$	Internal room volume, m^3

c_a = Specific heat capacity of air, $J/(kg\text{-}°C)$

C_z = Thermal capacitance of the zone, J/C

For modelling the total heat transfer for a single zonal room, we have to determine the various moisture sources effect on the moisture content in the air of the room. The source of the moisture are certain processes or equipment that produce moisture, such as cooking, foodstuffs, liquid water exposed to the air, etc. Other sources for moisture are occupants, and moisture from the outside air carried in the room through ventilation or infiltration. Unlike sensible heat gain, the latent heat gain of a room will instantaneously become a cooling load component. Therefore, we need to calculate the mass balance equation of a room or zone for the proper modelling of the heat transfer model of a single zonal room.

The mass balance equation for the zone is given by:

$$V \frac{dW_{bs}}{dt} = f_{sa}\left(W_{sa} - W_{bs}\right) + \frac{P(t)}{\rho_a} \tag{2.25}$$

where

V = Volume of the zone, m^3

W_{bs} = Humidity ratio of the building space, (dry air) kg/kg

W_{sa} = Humidity ratio of the supply, (dry air) kg/kg

$P(t)$ = Evaporation rate of the occupants, kg/h

2.4 STATE SPACE APPROACH FOR REPRESENTING A BUILDING ENERGY SYSTEM MODEL

In this model, Q_{HVAC} is the heat gain, which is supplied to the zone from the HVAC system in such a way that it gives the desired temperature in the room and also gives comfort to the occupant. Differential equations for building construction element and space can be represented in state space form as shown below.

$$\dot{X} = AX + BU \tag{2.26}$$
$$Y = CX + DU \tag{2.27}$$

where state vector (X) represents the state variables, which represent the room construction element temperatures, input vector (U) represents the input to the system, and the output of the system is the inner building space temperature (T_{bs}) and the building space humidity ratio (W_{bs}).

With the help of energy balance equations, we can represent the 3R2C thermal network model in the state space equation and output equations as follows.

$$
\begin{bmatrix}
\dot{T}_{w,o} \\
\dot{T}_{w,i} \\
\dot{T}_{w2,o} \\
\dot{T}_{w2,i} \\
\dot{T}_{g} \\
\dot{T}_{c,o} \\
\dot{T}_{c,i} \\
\dot{T}_{p,o} \\
\dot{T}_{p,i} \\
\dot{T}_{p2,o} \\
\dot{T}_{p2,i} \\
\dot{T}_{F,o} \\
\dot{T}_{F,i} \\
\dot{T}_{bs} \\
\dot{W}_{bs}
\end{bmatrix}
= A
\begin{bmatrix}
T_{w,o} \\
T_{w,i} \\
T_{w2,o} \\
T_{w2,i} \\
T_{g} \\
T_{c,o} \\
T_{c,i} \\
T_{p,o} \\
T_{p,i} \\
T_{p2,o} \\
T_{p2,i} \\
T_{F,o} \\
T_{F,i} \\
T_{bs} \\
W_{bs}
\end{bmatrix}
+ B
\begin{bmatrix}
T_{out} \\
Q_{ihg} \\
Q_{HVAC} \\
T_{bs1} \\
W_{sa} \\
P(t)
\end{bmatrix}
\tag{2.28}
$$

$$
Y = C
\begin{bmatrix}
T_{w,o} \\
T_{w,i} \\
T_{w2,o} \\
T_{w2,i} \\
T_{g} \\
T_{c,o} \\
T_{c,i} \\
T_{P,o} \\
T_{P,i} \\
T_{P2,o} \\
T_{P2,i} \\
T_{F,o} \\
T_{F,i} \\
T_{bs} \\
W_{bs}
\end{bmatrix}
+ D
\begin{bmatrix}
T_{out} \\
Q_{ihg} \\
Q_{HVAC} \\
T_{bs1} \\
W_{sa} \\
P(t)
\end{bmatrix}
\tag{2.29}
$$

The evaluated A, B, C, and D matrices are as follows.

$$
\begin{bmatrix}
A_{11} & A_{12} & 0 & 0 & 0 & 0 & 0 & 0 & 0 & 0 & 0 & 0 & 0 & 0 & 0 \\
A_{21} & A_{22} & 0 & 0 & 0 & 0 & 0 & 0 & 0 & 0 & 0 & 0 & 0 & A_{214} & 0 \\
0 & 0 & A_{33} & A_{34} & 0 & 0 & 0 & 0 & 0 & 0 & 0 & 0 & 0 & 0 & 0 \\
0 & 0 & A_{43} & A_{44} & 0 & 0 & 0 & 0 & 0 & 0 & 0 & 0 & 0 & A_{414} & 0 \\
0 & 0 & 0 & 0 & A_{55} & 0 & 0 & 0 & 0 & 0 & 0 & 0 & 0 & A_{514} & 0 \\
0 & 0 & 0 & 0 & 0 & A_{66} & A_{67} & 0 & 0 & 0 & 0 & 0 & 0 & 0 & 0 \\
0 & 0 & 0 & 0 & 0 & A_{76} & A_{77} & 0 & 0 & 0 & 0 & 0 & 0 & A_{714} & 0 \\
0 & 0 & 0 & 0 & 0 & 0 & 0 & A_{88} & A_{89} & 0 & 0 & 0 & 0 & 0 & 0 \\
0 & 0 & 0 & 0 & 0 & 0 & 0 & A_{98} & A_{99} & 0 & 0 & 0 & 0 & A_{914} & 0 \\
0 & 0 & 0 & 0 & 0 & 0 & 0 & 0 & 0 & A_{1010} & A_{1011} & 0 & 0 & 0 & 0 \\
0 & 0 & 0 & 0 & 0 & 0 & 0 & 0 & 0 & A_{1110} & A_{1111} & 0 & 0 & A_{1114} & 0 \\
0 & 0 & 0 & 0 & 0 & 0 & 0 & 0 & 0 & 0 & 0 & A_{1212} & A_{1213} & 0 & 0 \\
0 & 0 & 0 & 0 & 0 & 0 & 0 & 0 & 0 & 0 & 0 & A_{1312} & A_{1313} & A_{1314} & 0 \\
0 & A_{142} & 0 & A_{144} & A_{145} & 0 & A_{147} & 0 & A_{149} & 0 & A_{1411} & 0 & A_{1413} & A_{1414} & 0 \\
0 & 0 & 0 & 0 & 0 & 0 & 0 & 0 & 0 & 0 & 0 & 0 & 0 & 0 & A_{1515}
\end{bmatrix}
$$

where

$$A_{11} = -\left(\frac{1}{R_{w1}} + \frac{1}{R_{w2}}\right)\frac{A_{w1}}{C_{w1}} \qquad A_{12} = \frac{A_{w1}}{C_{w1}R_{w2}}$$

$$A_{21} = \frac{A_{w1}}{R_{w2}C_{w2}} \qquad A_{22} = -\left(\frac{1}{R_{w2}} + \frac{1}{R_{w3}}\right)\frac{A_{w1}}{C_{w2}}$$

$$A_{214} = \frac{A_{w1}}{R_{w3}C_{w2}}$$

$$A_{33} = -\left(\frac{1}{R_{w21}} + \frac{1}{R_{w22}}\right)\frac{A_{w2}}{C_{w21}} \qquad A_{34} = \frac{A_{w2}}{C_{w21}R_{w22}}$$

$$A_{43} = \frac{A_{w2}}{R_{w22}C_{w22}} \qquad A_{44} = -\left(\frac{1}{R_{w22}} + \frac{1}{R_{w23}}\right)\frac{A_{w2}}{C_{w22}}$$

$$A_{414} = \frac{A_{w2}}{R_{w23}C_{w22}}$$

$$A_{55} = -\left(\frac{1}{R_{g1}} + \frac{1}{R_{g2}}\right)\frac{A_g}{C_g} \qquad A_{514} = \frac{A_g}{C_g R_{g2}}$$

$$A_{66} = -\left(\frac{1}{R_{c1}} + \frac{1}{R_{c2}}\right)\frac{A_c}{C_{c1}} \qquad A_{67} = \frac{A_c}{C_{c1}R_{c2}}$$

$$A_{76} = \frac{A_c}{R_{c2}C_{c2}} \qquad A_{77} = -\left(\frac{1}{R_{c2}} + \frac{1}{R_{c3}}\right)\frac{A_c}{C_{c2}}$$

$$A_{714} = \frac{A_c}{R_{c3}C_{c2}}$$

$$A_{88} = -\left(\frac{1}{R_{P1}} + \frac{1}{R_{P2}}\right)\frac{A_P}{C_{P1}} \qquad A_{89} = \frac{A_P}{C_{P1}R_{P2}}$$

$$A_{98} = \frac{A_P}{R_{P2}C_{P2}} \qquad A_{99} = -\left(\frac{1}{R_{P2}} + \frac{1}{R_{P3}}\right)\frac{A_P}{C_{P2}}$$

$$A_{914} = \frac{A_P}{R_{P3}C_{P2}}$$

$$A_{1010} = -\left(\frac{1}{R_{P21}} + \frac{1}{R_{P22}}\right)\frac{A_{P2}}{C_{P21}} \qquad A_{1011} = \frac{A_{P2}}{C_{P21}R_{P22}}$$

$$A_{1110} = \frac{A_{P2}}{R_{P22}C_{P22}} \qquad A_{1111} = -\left(\frac{1}{R_{P22}} + \frac{1}{R_{P23}}\right)\frac{A_{P2}}{C_{P22}}$$

$$A_{1114} = \frac{A_{P2}}{R_{P23}C_{P22}}$$

$$A_{1212} = -\left(\frac{1}{R_{F1}} + \frac{1}{R_{F2}}\right)\frac{A_F}{C_{F1}} \qquad A_{1213} = \frac{A_F}{C_{F1}R_{F2}}$$

$$A_{1312} = \frac{A_F}{R_{F2}C_{F2}} \qquad A_{1312} = -\left(\frac{1}{R_{F2}} + \frac{1}{R_{F3}}\right)\frac{A_F}{C_{F2}}$$

$$A_{1314} = \frac{A_F}{R_{F3}C_{F2}}$$

$$A_{142} = \frac{A_{w1}}{R_{w3}\rho_a Vc_a C_z} \qquad A_{144} = \frac{A_{w2}}{R_{w23}\rho_a Vc_a C_z}$$

$$A_{145} = \frac{A_g}{R_{g2}\rho_a Vc_a C_z} \qquad A_{147} = \frac{A_c}{R_{c3}\rho_a Vc_a C_z}$$

$$A_{149} = \frac{A_p}{R_{P3}\rho_a Vc_a C_z} \qquad A_{1411} = \frac{A_{P2}}{R_{P23}\rho_a Vc_a C_z}$$

$$A_{1413} = \frac{A_F}{R_{F3}\rho_a Vc_a C_z} \qquad A_{1414} = -\left(\frac{A_{w1}}{R_{w3}} + \frac{A_{w2}}{R_{w23}} + \frac{A_g}{R_{g2}} + \frac{A_c}{R_{c3}}\right.$$

$$\left. + \frac{A_P}{R_{P3}} + \frac{A_{P2}}{R_{P23}} + \frac{A_F}{R_{F3}}\right)\frac{1}{\rho_a Vc_a C_z}$$

$$A_{1515} = \frac{-f_{sa}}{V}$$

$$
\begin{bmatrix}
\dfrac{A_{w1}}{R_{w1}C_{w1}} & 0 & 0 & 0 & 0 & 0 \\[2ex]
0 & \dfrac{A_{w1}}{C_{w2}} & \dfrac{A_{w1}}{C_{w2}} & 0 & 0 & 0 \\[2ex]
\dfrac{A_{w2}}{R_{w21}C_{w21}} & 0 & 0 & 0 & 0 & 0 \\[2ex]
0 & \dfrac{A_{w2}}{C_{w22}} & \dfrac{A_{w2}}{C_{w22}} & 0 & 0 & 0 \\[2ex]
\dfrac{A_g}{R_{g1}C_g} & \dfrac{A_g}{C_g} & \dfrac{A_g}{C_g} & 0 & 0 & 0 \\[2ex]
\dfrac{A_c}{R_{c1}C_{c1}} & 0 & 0 & 0 & 0 & 0 \\[2ex]
0 & \dfrac{A_c}{C_{c2}} & \dfrac{A_c}{C_{c2}} & 0 & 0 & 0 \\[2ex]
0 & 0 & 0 & \dfrac{A_P}{R_{P1}C_{P1}} & 0 & 0 \\[2ex]
0 & \dfrac{A_P}{C_{P2}} & \dfrac{A_P}{C_{P2}} & 0 & 0 & 0 \\[2ex]
0 & 0 & 0 & \dfrac{A_{P2}}{R_{P21}C_{P21}} & 0 & 0 \\[2ex]
0 & \dfrac{A_P}{C_{P22}} & \dfrac{A_P}{C_{P22}} & 0 & 0 & 0 \\[2ex]
0 & 0 & 0 & \dfrac{A_F}{R_{F1}C_{F1}} & 0 & 0 \\[2ex]
0 & \dfrac{A_F}{C_{F2}} & \dfrac{A_F}{C_{F2}} & 0 & 0 & 0 \\[2ex]
0 & \dfrac{1}{\rho_a V c_a C_Z} & \dfrac{1}{\rho_a V c_a C_Z} & 0 & 0 & 0 \\[2ex]
0 & 0 & 0 & 0 & \dfrac{f_{sa}}{V} & \dfrac{1}{V\rho_a}
\end{bmatrix}
$$

C matrix:

$$
\begin{bmatrix}
0 & 0 & 0 & 0 & 0 & 0 & 0 & 0 & 0 & 0 & 0 & 0 & 0 & 1 & 0 \\
0 & 0 & 0 & 0 & 0 & 0 & 0 & 0 & 0 & 0 & 0 & 0 & 0 & 0 & 1
\end{bmatrix}
$$

All the elements of the D matrix are zero.

2.5 DYNAMIC RESPONSE FOR THE DEVELOPED CONSTRUCTION ELEMENT AND BUILDING SPACE MODEL

The transient response characteristics of a system is specified by the following:

1. **Delay time**, t_d: The time required for the response to reach half the final value the very first time.
2. **Rise time** t_r: The time required for the response to rise from 10% to 90%, 5% to 95%, or 0% to 100% of its final value.
3. **Peak time**, t_p: The time required for the response to reach the first peak of the overshoot.
4. **Maximum overshoot**, Mp: The maximum peak value of the response if different from unity. If the final steady state value of the response differs from unity, then use the maximum percent overshoot, which is defined as:

$$\text{Maximum percent overshoot} = \frac{c(t_p) - c(\infty)}{c(\infty)} \times 100\%$$

5. **Settling time**, t_s: The time taken for the response of the system to reach and settle to the limits of the tolerance band (either 2% or 5%).

2.5.1 Step response for the external walls

Figure 2.2 to Figure 2.4 show the step response of external wall 1 and external wall 2 when only one input (step input) is given to the system at a time and the remaining two are input as zero.

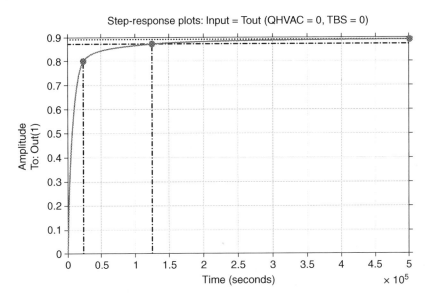

Figure 2.2 Step response of external wall 1 and external wall 2, when input is T_{out}.

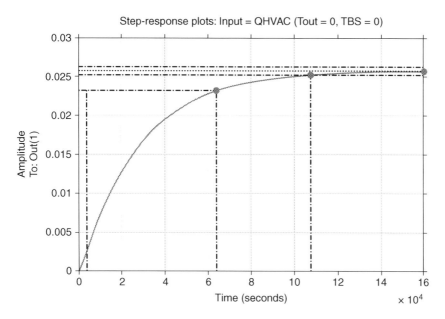

Figure 2.3 Step response of external wall 1 and external wall 2, when input is Q_{HVAC}.

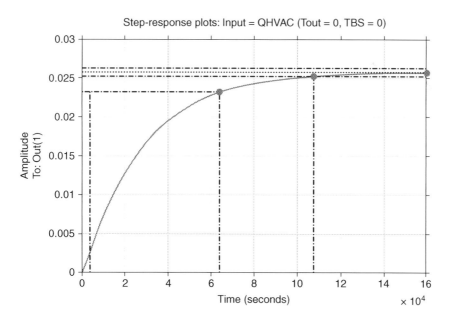

Figure 2.4 Step response of external wall 1 and external wall 2, when input is T_{BS}.

Table 2.1 Performance characteristics of external wall 1 and external wall 2

Characteristics	$G_1(s)$	$G_2(s)$	$G_3(s)$
Rise time (sec)	23000	59800	59800
Settling time (sec)	125000	108000	108000
Peak value	0.89	0.257	7.41×10^9

The rise time, settling time, and peak value for all three conditions are shown in Table 2.1.

2.5.2 Step response for the partition walls

Figure 2.5 to Figure 2.7 show the step response of partition wall 1 and partition wall 2 when only one input (step input) is given to the system at a time and the remaining two are input as zero. The rise time, settling time, and peak value for all three conditions are shown in Table 2.2.

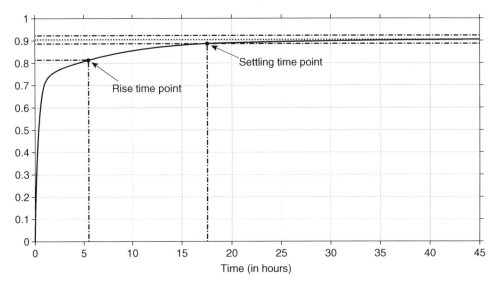

Figure 2.5 Step response of partition wall 1 and partition wall 2, when input is T_{out}.

Table 2.2 Performance characteristics of partition wall 1 and partition wall 2

Characteristics	$G_1(s)$	$G_2(s)$	$G_3(s)$
Rise time (sec)	28200	353000	353000
Settling time (sec)	280000	629000	629000
Peak value	0.892	0.528	4.9×10^{10}

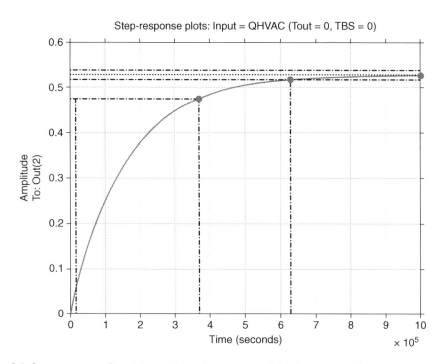

Figure 2.6 Step response of partition wall 1 and partition wall 2, when input is Q_{HVAC}.

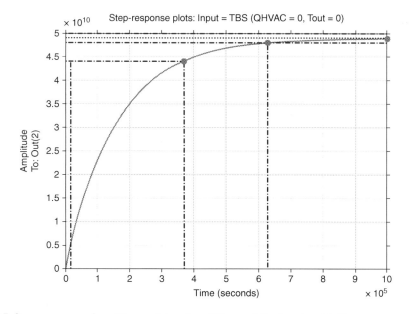

Figure 2.7 Step response of partition wall 1 and partition wall 2, when input is T_{BS}.

2.5.3 Step response for ceiling and roof

Figure 2.8 to Figure 2.10 show the step response of the ceiling and roof when only one input (step input) is given to the system at a time and the remaining two are input as zero.

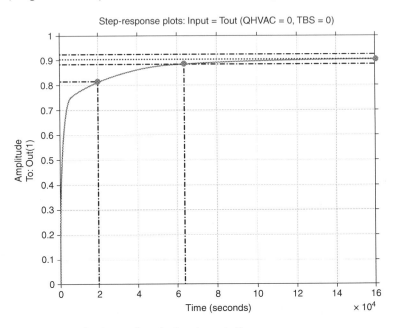

Figure 2.8 Step response of ceiling and roof, when input is T_{out}.

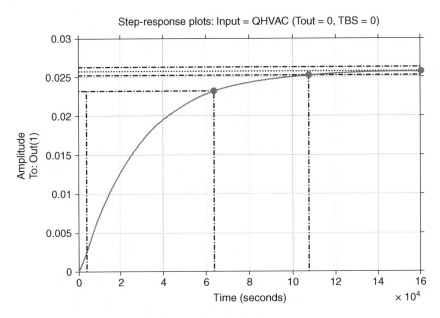

Figure 2.9 Step response of ceiling and roof, when input is Q_{HVAC}.

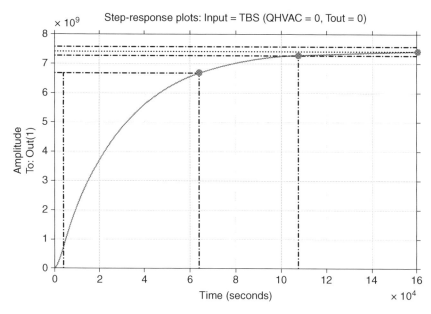

Figure 2.10 Step response of ceiling and roof, when input is T_{BS}.

Table 2.3 Performance characteristics of ceiling and roof

Characteristics	$G_1(s)$	$G_2(s)$	$G_3(s)$
Rise time (sec)	19600	598000	598000
Settling time (sec)	63500	108000	108000
Peak value	0.905	0.0257	7.41×10^9

The rise time, settling time, and peak value for all three conditions are shown in Table 2.3.

2.5.4 Step response for the building space

In MATLAB, the unit step input was given for the state space model of a 3R2C thermal network model of a room, and the following results were found. Figure 2.11 shows the step response of the inside room temperature (T_{BS}) when the step outdoor temperature (T_{out}) as input (1) is given to the model and the remaining two inputs are zero. Figure 2.11 also shows that the inside room temperature rises in approximately 3470 seconds and settles at a final value of 0.541 with an approximate settling time of 8300 seconds.

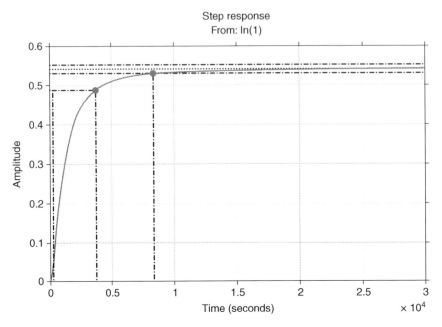

Figure 2.11 Step response of building space air temperature at input (1).

Figure 2.12 shows the step response of the inside room temperature (T_{BS}) when the step HVAC heat gain (Q_{HVAC}) as input (2) is given to the model and the remaining two inputs are zero. The indoor room temperature rises in approximately 5960 seconds and settles at a final value of 0.312 with an approximate settling time of 1270 seconds.

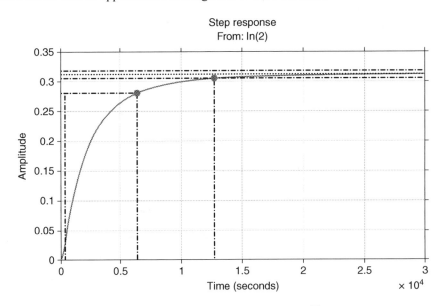

Figure 2.12 Step response of building space air temperature at input (2).

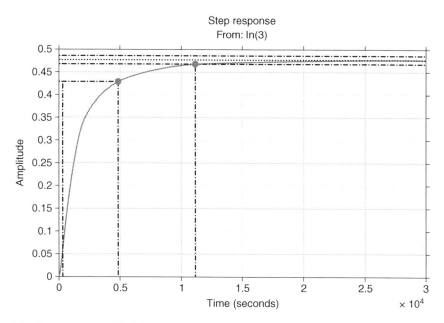

Figure 2.13 Step response of building space air temperature at input (3).

Figure 2.13 shows the step response of the inside room temperature (T_{BS}) when the step building space air temperature 1 (T_{BS1}, i.e., the temperature of the adjacent room) as input (3) is given to the model and the remaining two inputs are zero. The indoor room temperature rises in approximately 4520 seconds and settles at a final value of 0.476 with an approximate settling time of 1120 seconds. The simulation result shows that the rise time is very high when the step input is given to the model, and in Figure 2.11, when the input as a step outdoor temperature is given to the model, the settling time is severely high.

REFERENCES

Harish, V. S. K. V., & Kumar, A. (2014, January). Techniques used to construct an energy model for attaining energy efficiency in building: A review. In *Proceedings of The 2014 International Conference on Control, Instrumentation, Energy and Communication (CIEC)* (pp. 366–370). IEEE, 10.1109/CIEC.2014.6959111.

Harish, V. S. K. V., & Kumar, A. (2016). A review on modeling and simulation of building energy systems. *Renewable and Sustainable Energy Reviews*, 56, 1272–1292, doi.org/10.1016/j.rser.2015.12.040.

Harish, V. S. K. V., & Kumar, A. (2016). Reduced order modeling and parameter identification of a building energy system model through an optimization routine. *Applied Energy*, 162, 1010–1023, doi.org/10.1016/j.apenergy.2015.10.137.

Harish, V. S. K. V., & Kumar, A. (2016, January). Modeling and simulation of a simple building energy system. In *2016 International Conference on Microelectronics, Computing and Communications (MicroCom)* (pp. 1–6). IEEE, doi.org/10.1109/MicroCom.2016.7522473.

Harish, V. S. K. V., & Kumar, A. (2019, October). Stability analysis of reduced order building energy models for optimal energy control. In *2019 2nd International Conference on Power Energy, Environment and Intelligent Control (PEEIC)* (pp. 327–331). IEEE, doi.org/10.1109/PEEIC47157.2019.8976699.

Harish, V. S. K. V., Sant, A. V., & Kumar, A. (2021). Determining the performance characteristics of a white-box building energy system model and evaluating the energy consumption. In *Advances in Clean Energy Technologies* (pp. 605–615). Springer, Singapore, doi.org/10.1007/978-981-16-0235-1_46.

Underwood, C., & Yik, F. (2008). *Modelling methods for energy in buildings*. John Wiley & Sons, doi.org/10.1002/9780470758533.

Chapter 3

Modelling and simulation of heating ventilation and air-conditioning system

V.S.K.V. Harish, Nayan Kumar Singh, Arun Kumar, Karan Doshi, and Amit Vilas Sant

CONTENTS

3.1 INTRODUCTION

The heating, ventilation, and air conditioning (HVAC) system has a large impact on a building's total energy consumption. It consumes 40% to 60% of the total electricity of a commercial building. In an HVAC system, the most energy is consumed by the compressors, which is 61% of the total energy consumption of the HVAC. The second-highest energy is consumed by the air handling unit (AHU), which represents 13% of the total energy consumption of the HVAC (Harish & Kumar, 2016). The chilled water pumps and condenser water pumps consume 11% and 7%, respectively, and the fan coil unit and cooling towers consume up to 5% and 3%, respectively, of the total energy consumption of the HVAC, as shown in Figure 3.1.

The primary goal of any HVAC system is to ensure the comfort of the occupants. Air quality is an essential consideration for maintaining the occupants' productivity, comfort, and health and should not be underestimated. At a very intuitive level, an HVAC system has five loops:

DOI: 10.1201/9781003211587-3

1. Airside loop
2. Chilled water loop
3. Refrigeration loop
4. Heat rejection loop
5. Control loop

These loops can be used to define any HVAC system virtually, but it is not necessary for every HVAC to use these five loops.

In the airside loop, in order to maintain the dry bulb temperature and the humidity level of the conditioned space, heat and moisture must be added or removed at the same rate as it leaves or enters the conditioned space. The supply air is cooled and dehumidified by a cooling coil and then sent to the space. It should be cold enough to absorb the extra sensible heat from the space and dry enough to absorb the extra moisture.

The second coil is the chilled water loop, as the cooling coil is used to cool and dehumidify the supply air. The fluid flowing through the cooling coil is the refrigerant agent and can be cold water or a liquid refrigerant. A system that uses water as the flowing liquid in the coil is known as a chilled water loop. Heat flows from a higher temperature region to a lower temperature region. Therefore, the heat will be transferred from the air to the fluid flowing in the cooling coil, which is much colder than the air.

The third loop is the refrigeration loop, where heat is transferred from the water to the refrigerant, the liquid refrigerant boils, and the vapours of the refrigerant are heated more (superheated) inside the evaporator before being delivered to the compressor.

The fourth loop is the heat rejection loop. In a water-cooled condenser, the cooled water flows inside the tubes and the refrigerant vapours pass through the coils and transfer the heat from the refrigerant vapours to the water that flows in the condenser coil. When a

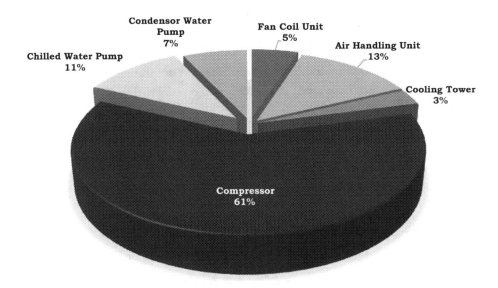

Figure 3.1 Energy consumption in an HVAC system.

water-cooled condenser is used, the heat exchanger is normally either a fluid cooler or cooling tower.

The last loop is the control loop, which controls the components of the HVAC system so that it can operate efficiently and save energy while maintaining human comfort.

HVAC systems generate heat. The air system takes this heat and passes it to the water system, which passes it to the refrigeration system. The refrigeration system takes the heat and passes it to the cooling tower, which ultimately gives it to the environment. This process is the goal of any HVAC system. There are many types of HVAC systems, but we are going to mostly focus on the forced air HVAC system. The forced air comes from the fact that we are literally pushing cool air/hot air into the room during summer/winter time. In the HVAC system model, the major components are the AHU and an air-conditioned single zonal room. The AHU consists of a mixing box, heating coil, humidifier, cooling coil and dehumidifier, fan, and ductwork.

3.2 DEVELOPMENT OF THE DYNAMIC MODEL OF AN HVAC SYSTEM

3.2.1 HVAC system description

In the HVAC system model, the major components are the air handling unit (AHU) and an air-conditioned single zonal room. The AHU consists of a mixing box, heating coil, humidifier, cooling coil and dehumidifier, fan, and ductwork, as shown in Figure 3.2. During winter season operation, the heating system model will come into play. The heating system model consists of a mixing box, heating coil, humidifier, fan, ductwork, and air-conditioned single zonal room.

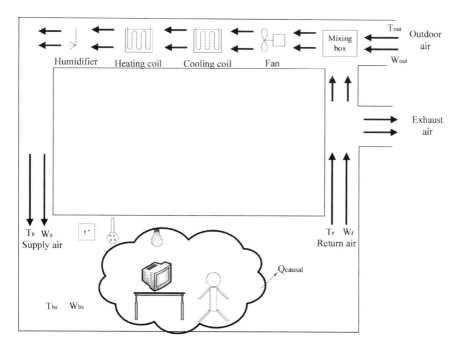

Figure 3.2 Schematic diagram of an HVAC system.

During winter season operation, cold and dry air enters the AHU from the outside. The thermostat will sense the temperature in the single zonal room and send the signal to the controller, which uses the error signal to control the input temperature of the water that flows through the heating coil. Finally, the air passes through the humidifier, which generates vapour to control the humidity ratio in the air that is supplied to the single zonal room.

We assume that both the temperature of the air that is returned from the single zonal room through the ducts and the zone temperature is the same. This can be represented as:

$$Q_{HVAC} = m_{sa} c_a \left(T_{sa} - T_{bs} \right) \tag{3.1}$$

$$m_{sa} = f_{sa} \, \rho_a$$

where

m_{sa} = Mass flow rate of the supply air, kg/s

f_{sa} = Volume flow rate of the supply air, m^3/s

$\dot{A}_a$ = Density of air, kg/m^3

T_{sa} = Supply air temperature, $°C$

T_{bs} = Building space temperature, $°C$

From the above equation, we can see that there are only two control inputs, mass flow rate (m_{sa}) and supply temperature (T_{sa}), for controlling the zone temperature. And in this model of HVAC system, we assume the mass flow rate (m_{sa}) is constant and we will control the supply temperature (T_{sa}).

Figure 3.3 shows the block diagram of the building energy model with an HVAC system, in which the output of the building energy model is the building space temperature (T_{bs}) and the humidity ratio of the zone (W_{bs}). The sensor will measure the temperature and humidity in the building space and send it to the summing block. The error between the reference temperature and the reference humidity ratio is given as the input to the PID controller block. Then PID controller block will send the control signal to the HVAC system in such a way that it will control the temperature and humidity ratio inside the building space as per the reference.

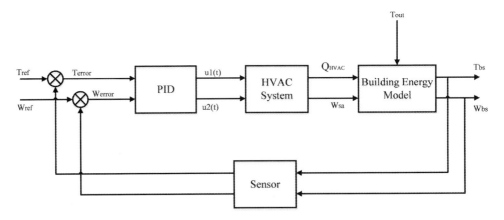

Figure 3.3 Block diagram of a building energy model with an HVAC system.

3.2.1.1 Heating coil model

The heating coil in an HVAC system is a water-to-air heat exchanger that provides conditioned air to the zone for ventilation purposes. For modelling the heating coil model, the following assumptions are made:

1. The mass flow rate of the water inside the coil is constant.
2. The coil material is highly conductive such that its thermal resistance is considered to be negligible.

In the heating coil model, the heated water is supplied to the inlet of the coil at temperature T_{wi} and the temperature of the water at the outlet of the coil is T_{wo}, which is assumed to be constant.

The energy balance equation between hot water and cold air can be written as:

$$C_{ahu}\frac{dT_{ao}}{dt} = f_{sw}\rho_w c_w \left(T_{wi} - T_{wo}\right) + \left(UA\right)_{ahu}\left(T_o - T_{ao}\right) + f_{sa}\rho_a c_a \left(T_m - T_{ao}\right) \qquad (3.2)$$

where

C_{ahu}	=	Overall thermal capacitance of the air handling unit, $kJ/°C$
f_{sa}	=	Volume flow rate of water, m^3/s
f_{sw}	=	Volume flow rate of the supply air, m^3/s
$(UA)_{ahu}$	=	Overall transmittance area factor of the air handling unit, $kJ/s°C$
T_{ao}	=	Temperature of the air out from the coil, $°C$
ρ_a	=	Density of air, kg/m^3
ρ_w	=	Density of water, kg/m^3
c_w	=	Specific heat capacity of water, $J/(kg\text{-}°C)$
T_{wi}	=	Supply water temperature, $°C$
T_{wo}	=	Return water temperature, $°C$
T_m	=	Temperature of the air out from the mixing box, $°C$

The mass balance equation of a heating coil is given by:

$$V_{ahu}\frac{dW_{ao}}{dt} = f_{sa}\left(W_{sa} - W_{ao}\right) \qquad (3.3)$$

where

V_{ahu}	=	Volume of the air handling unit, m^3
W_{ao}	=	Humidity ratio of the air out from the coil, (dry air) kg/kg
W_{sa}	=	Humidity ratio of the supply, (dry air) kg/kg

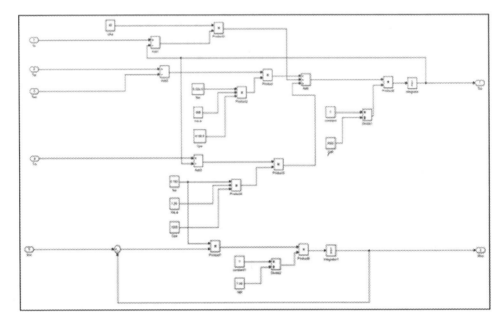

Figure 3.4 Simulink diagram of a heating coil model.

3.2.1.2 Humidifier model

Humidification is a mass transfer process of water vapours in the atmospheric air.

Due to this process, the quantity of water vapours increases in the mixture. There will be an undesirable effect on the human body when the moisture content inside the zone is extremely low. Therefore, controlling the moisture content inside the zone is important for air conditioning and occupant comfort.

The energy balance equation for the humidifier model can be written as:

$$C_h \frac{dT_h}{dt} = f_{sa}c_a\left(T_{si} - T_h\right) + \left(UA\right)_h\left(T_o - T_h\right) \tag{3.4}$$

where

C_h	=	Overall thermal capacitance of the humidifier, $kJ/°C$
f_{sa}	=	Volume flow rate of the supply air, m^3/s
$(UA)_h$	=	Overall transmittance area factor of the humidifier, $kJ/s°C$
T_h	=	Temperature of the supply air after passing humidifier, $°C$
T_{si}	=	Temperature of the supply air (to the humidifier), $°C$
T_o	=	Outdoor temperature, $°C$

The mass balance equation for the humidifier model is given by:

$$V_h \frac{dW_h}{dt} = f_{sa}\left(W_{si} - W_h\right) + \frac{h(t)}{\rho_a}$$

(3.5)

where

V_h =   Volume of the humidifier, m^3

W_h =   Humidity ratio of the supply air out of the humidifier, (dry air) kg/kg

W_{si} =   Humidity ratio of the supply to the humidifier, (dry air) kg/kg

$h(t)$ =   Rate of moisture air produced in the humidifier

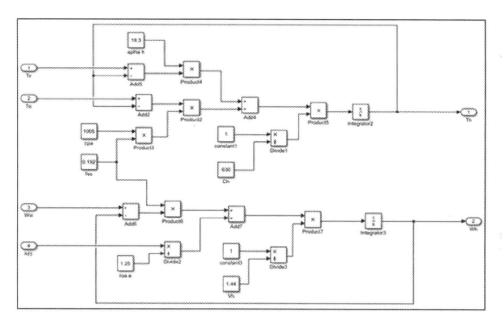

Figure 3.5 Simulink diagram of a humidifier model.

3.2.1.3 Fan model

The fan of an HVAC system is represented by the first-order differential equation. It is assumed that the air temperature changes have a negligible effect on the physical properties of the air. There is no effect on the humidity ratio because no mass transfer takes place.

3.2.1.4 Mixing box model

In the mixing box model, the recirculated air from the zone and the outdoor air are mixed together. The air supplied to the heating coil/cooling coil is comprised of a fraction of the returned air from the zone along with the outdoor air.

The energy balance equation for the mixing box model is given by:

$$m_r c_a T_r + m_0 c_a T_{out} = m_m c_a T_m$$

$$m_r + m_{out} = m_m$$

$$T_m = \frac{m_r T_r + m_0 T_{out}}{m_r + m_{out}}$$
(3.6)

where

T_m	=	Temperature of the air out of the mixing box, °C
m_r	=	Mass flow rate of the recalculated air, kg/s
m_{out}	=	Mass flow rate of the outdoor air, kg/s
T_{out}	=	Outdoor temperature, °C

The mass balance equation for the mixing box model is given by:

$$W_m = \frac{m_r W_r + m_{out} W_{out}}{m_r + m_{out}}$$
(3.7)

where

W_m	=	Humidity ratio of air out of the mixing box, (dry air) kg/kg
W_r	=	Humidity ratio of the recirculated air, (dry air) kg/kg
W_{out}	=	Outdoor humidity ratio, (dry air) kg/kg

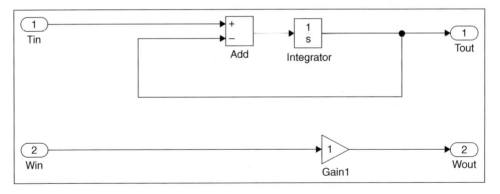

Figure 3.6 Simulink diagram of a fan model.

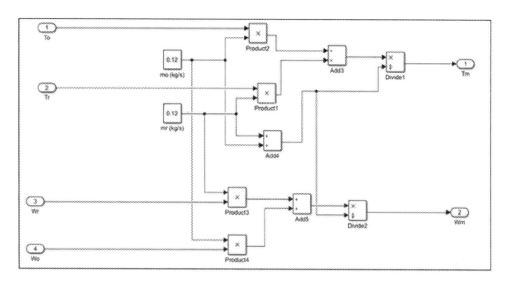

Figure 3.7 Simulink diagram of a mixing box model.

3.2.1.5 Duct model

The duct unit is represented by the transient model. The energy balance equation for the duct is given by:

$$\frac{dT_{d,0}}{dt} = \frac{(h_i + h_0)m_a c_a}{h_i M_d C_d}(T_{in} - T_{d,0})$$ (3.8)

where

$T_{d,0}$	=	Temperature of the air outside the duct, $°C$
T_{in}	=	Temperature of the air in the duct, $°C$
h_i	=	Heat transfer coefficient inside the duct, $W/m^2 \, °C$
m_a	=	Mass flow rate of the supplied air, kg/s
h_0	=	Heat transfer coefficient in the ambient, $W/m^2 \, °C$
M_d	=	Mass of the duct model, kg/m
C_d	=	Specific heat of the duct model, $kJ/kg \, °C$

There is no effect on the humidity ratio.

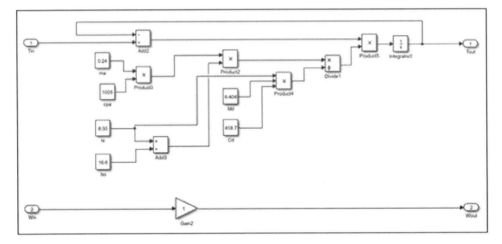

Figure 3.8 Simulink diagram of a duct model.

3.2.1.6 Sensor model

In this model, sensors are used to measure the inside building space temperature and humidity ratio and give a feedback signal to the controller in order to enhance the performance of the system. For simplicity, the sensor is considered as a first-order differential equation.

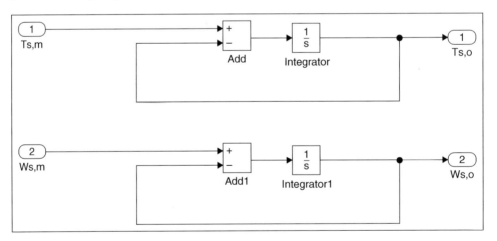

Figure 3.9 Simulink diagram of a sensor model.

3.3 BUILDING ENERGY SYSTEM MODEL ANALYSIS AND EVALUATION

The dynamic energy transfer equations involved in the development of a building energy system model are time and space dependent. For analyzing the energy transfer processes, a test case is used, as shown in Figure 3.10.

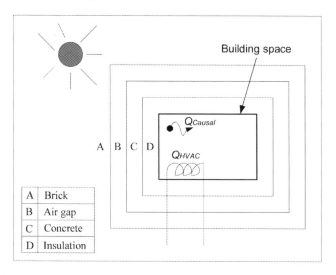

Figure 3.10 Building space with multilayered building construction elements (A, B, C, D).

Simulations are performed for a complete building space whose building construction elements are those as given in Tables 3.1 and 3.2.

Table 3.1 Thermophysical properties of the materials making up the building walls

Building element	Wall layers	Thickness (m)	Thermal conductivity (W/m °C)	Density (kg/m³)	Specific heat capacity (J/kg °C)
Outdoor wall	Insulation	0.12	0.04	30	840
	Air gap	0.05	0.56	1	1000
	Brick	0.10	0.60	1500	840
Adjacent wall (1)	Brick	0.122	0.84	800	1700
	Insulation	0.05	0.03	1764	30
	Concrete block	0.112	0.510	1000	1400
	Plaster	0.013	0.160	1000	600
Adjacent wall (2)	Aluminium	0.003	0.160	896	2800
	Air gap	0.100	–	1025	1.20
	Insulation	0.075	0.035	1000	30
	Cast concrete	0.185	1.130	1000	2000
	Plaster	0.013	0.160	1000	600
Roof and floor	Carpet	0.009	0.060	2500	160
	Screed	0.065	0.410	840	1200
	Concrete	0.125	1.130	1000	2000
Lobby wall	Plaster	0.013	0.160	1000	600
	Concrete block	0.122	0.510	1000	1400
	Plaster	0.013	0.160	1000	600

The numerical values of the building energy system model elements, such as building space, HVAC, and lighting system, are given in Table 3.2.

Table 3.2 Numerical values of the building energy system model parameters

Building energy system model	Parameter values	
Building space/zonal characteristics		
Building envelope	Volume (m³)	500
	Outdoor wall area (m²)	60
	Outdoor wall thermal resistance (m² °C/W)	3.2560
	Outdoor wall thermal capacitance (J/m² °C)	129074
	Adjacent wall area (m²)	40
	Adjacent wall thermal resistance (m² °C/W)	2.1128
	Adjacent wall thermal capacitance (J/m² °C)	333166
	Roof/floor area (m²)	120
	Roof/floor thermal resistance (m² °C /W)	0.4192
	Roof/floor thermal capacitance (J/m² °C)	319120
	Lobby wall area (m²)	35
	Lobby wall thermal resistance (m² °C/W)	0.4017
	Lobby wall thermal capacitance (J/m² °C)	170772
Window	Window area (m²)	16
	Window U-value (W/m² °C)	2.8
	Glass/glazing area (m²)	8
	Glass transmitivity	75%
	Ground reflectance	25%
Occupancy schedule		
	Daily operating hours	8
	Start hours	10.00
	Weekly operating days	6
HVAC system		
Ventilation	Reference air changes per hour (h⁻¹)	0.6
Control valve	Designed mass flow rate (kg/s)	0.252
Heat exchanger	Heat transfer coefficient (W/m² °C)	32.2
	Thermal capacitance of fluid and material (J/°C)	68300
Lighting		
Type of luminaire	Pendant fluorescent 2440 mm type lamps	
	Number of lamps	8
	Wattage of each lamp (W)	8
	Special allowance factor (F_{SA})	0.85
	Lighting use factor (F_{use})	1

The different layers of the multilayered building construction element are shown in Figure 3.11.

x_1, x_2, x_3: the thickness of the layers (m)

A_1, A_2, A_3, A_4: the cross-sectional areas of the layers (m²)

T_{ins}, T_{conc}, T_b: the temperature of the insulation node, concrete, and brick, respectively (°C)

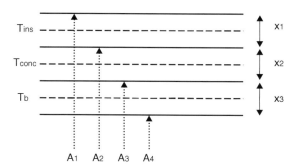

Figure 3.11 Multilayered building construction element parameters.

For the building space under study, the energy balance equation is given as equation (3.9).

$$\rho_a C_{p\_a} V_{bs} \frac{dT_{bs}}{dt} = \dot{Q}_{HVAC} + \dot{Q}_{casual} - \dot{V}_{bs} \rho_a C_{p\_a} \left(T_{bs} - T_{vent} \right) - \dots$$
$$\frac{1}{\dfrac{1}{h_i} + \dfrac{x_1}{2k_{ins}}} A_{ins} \left(T_{bs} - T_{ins} \right) - \frac{1}{\dfrac{1}{h_i} + \dfrac{x_{1,rr}}{2k_{ins,r}}} A_{ins,r} \left(T_{bs} - T_{ins,r} \right) \qquad (3.9)$$

Following similar principles, the energy balance equations for the other layers of the building construction elements making up the building space are deduced as equation (3.10) to equation (3.12).

$$\rho_{ins} C_{p\_ins} V_{ins} \frac{dT_{ins}}{dt} = \frac{1}{\dfrac{1}{h_i} + \dfrac{x_1}{2k_{ins}}} A_{ins} \left(T_{bs} - T_{ins} \right) - \dots$$
$$\frac{2k_{ins} k_{conc}}{x_1 k_{conc} + x_2 k_{ins}} A_{conc} \left(T_{ins} - T_{conc} \right) \qquad (3.10)$$

$$\rho_{conc} C_{p\_conc} V_{conc} \frac{dT_{conc}}{dt} = \frac{2k_{ins} k_{conc}}{x_1 k_{conc} + x_2 k_{ins}} A_{conc} \left(T_{ins} - T_{conc} \right) - \dots$$
$$\frac{2k_{conc} k_b}{x_2 k_b + x_3 k_{conc}} A_{conc} \left(T_{conc} - T_b \right) \qquad (3.11)$$

$$\rho_b C_{p\_b} V_b \frac{dT_b}{dt} = \frac{2k_b k_{conc}}{x_3 k_{conc} + x_2 k_b} A_b \left(T_{conc} - T_b \right) - \dots$$
$$\frac{1}{\dfrac{1}{h_o} + \dfrac{x_3}{2k_b}} A_{out} \left(T_b - T_{out} \right) \qquad (3.12)$$

Equation (3.9) to equation (3.12) is simplified in matrix form as equation (3.13) and equation (3.14).

$$
\begin{pmatrix} \dot{T}_{bs} \\ \dot{T}_{ins} \\ \dot{T}_{conc} \\ \dot{T}_{b} \end{pmatrix} = \begin{bmatrix} \dfrac{-\dot{V}_{bs}\rho_a C_{p\_a} - \omega_1 A_1}{\rho_a C_{p\_a} V_{bs}} & \dfrac{\omega_1 A_1}{\rho_a C_{p\_a} V_{bs}} & 0 & 0 \\[2ex] \dfrac{\omega_1 A_1}{\rho_{ins} C_{p\_ins} V_{ins}} & \dfrac{-\omega_1 A_1 - \omega_2 A_2}{\rho_{ins} C_{p\_ins} V_{ins}} & \dfrac{\omega_2 A_2}{\rho_{ins} C_{p\_ins} V_{ins}} & 0 \\[2ex] 0 & \dfrac{\omega_2 A_2}{\rho_{conc} C_{p\_conc} V_{conc}} & \dfrac{-\omega_1 A_1 - \omega_3 A_3}{\rho_{conc} C_{p\_conc} V_{conc}} & \dfrac{\omega_3 A_3}{\rho_{conc} C_{p\_conc} V_{conc}} \\[2ex] 0 & 0 & \dfrac{\omega_3 A_3}{\rho_b C_{p\_b} V_b} & \dfrac{-\omega_3 A_3 - \omega_4 A_4}{\rho_b C_{p\_b} V_b} \end{bmatrix} \cdots
$$

$$
\begin{pmatrix} T_{bs} \\ T_{ins} \\ T_{conc} \\ T_b \end{pmatrix} + \begin{bmatrix} \dfrac{1}{\rho_a C_{p\_a} V_{bs}} & 0 & \dfrac{\dot{V}_{bs}}{V_{bs}} \\[2ex] 0 & 0 & 0 \\[2ex] 0 & 0 & 0 \\[2ex] 0 & \dfrac{\omega_4 A_4}{\rho_b C_{p\_b} V_b} & 0 \end{bmatrix} \begin{pmatrix} \dot{Q}_{BS} \\ T_{out} \\ T_{bs} \end{pmatrix} \tag{3.13}
$$

and

$$
\left(T_{o/p} \right) = \begin{bmatrix} 1 & 0 & 0 & 0 \end{bmatrix} \begin{pmatrix} T_{bs} \\ T_{ins} \\ T_{conc} \\ T_b \end{pmatrix} + \begin{bmatrix} 0 & 0 & 0 \end{bmatrix} \begin{pmatrix} \dot{Q}_{BS} \\ T_{out} \\ T_{bs} \end{pmatrix} \tag{3.14}
$$

Also, the net heat rate ($\dot{Q}_{BS}$) in the building space due to the HVAC system ($\dot{Q}_{HVAC}$) and causal factors ($\dot{Q}_{casual}$) such as occupancy, furniture, electronic equipment, etc., is given as equation (3.15).

$$
\dot{Q}_{BS} = \dot{Q}_{HVAC} + \dot{Q}_{causal} \tag{3.15}
$$

Parameters (ω_1, ω_2, ω_3, and ω_4) used in equation (3.13) are defined as equation (3.16) to equation (3.19).

$$
\omega_1 = \dfrac{1}{\dfrac{1}{h_i} + \dfrac{x_1}{2k_{ins}}} \tag{3.16}
$$

$$
\omega_2 = \dfrac{2k_{ins}k_{conc}}{x_1 k_{conc} + x_2 k_{ins}} \tag{3.17}
$$

$$
\omega_3 = \dfrac{2k_b k_{conc}}{x_3 k_{conc} + x_2 k_b} \tag{3.18}
$$

$$\omega_4 = \frac{1}{\dfrac{x_3}{2k_b} + \dfrac{1}{h_o}} \tag{3.19}$$

Equations (3.13) and (3.14) are represented in state-space form as equation (3.20).

$$\begin{aligned} \dot{T} &= AT + BU \\ T_{o/p} &= CT + DU \end{aligned} \tag{3.20}$$

where

T	=	State vector
U	=	Vector of inputs to the state-space model
$T_{o \cdot p}$	=	Output vector of the state-space model
A, B, C, and D	=	Coefficient matrices

The transfer function for the developed state-space model is given as equation (3.21).

$$G(s) = C(sI - A)^{-1}B + D \tag{3.21}$$

where

S	=	Laplace transform variable
$G(s)$	=	System transfer function

The $|sI - A|$ term in equation (3.21) is the characteristic polynomial of $G(s)$, and the eigenvalues of A are the identical poles of $G(s)$. The transformation of a state-space model into a transfer function enables one to describe a dynamic relationship between the inputs or excitations and the outputs or responses of the system under study. The Laplace transformation is applied on input-output differential equations with a condition that all initial values are zero or changing infinitesimally.

The state-space model developed using equation (3.21) does not take into consideration the effect of windows or glazing area within a building space. Assuming that one-third (33%) of the inner space area is exposed to opaque elements, the developed building energy system is analyzed using equations (3.14) and (3.15). The effect of the windowpanes surrounding the glass is neglected as the windowpanes and frames do not account for appreciable heat transfer within a building space. For computing building space air temperature, the energy balance equation for the building energy system model under study is given as equation (3.22).

$$\begin{aligned} \rho_a C_{p\_a} V_{bs} \frac{dT_{bs}}{dt} &= \dot{Q}_{HVAC} + \dot{Q}_{casual} - \dot{V}_{bs}\rho_a C_{p\_a}(T_{bs} - T_{vent}) - \dots \\ &\quad \frac{1}{\dfrac{1}{h_i} + \dfrac{x_1}{2k_{ins}}} 0.67 A_{ins}(T_{bs} - T_{ins}) - \dots \\ &\quad \frac{1}{\dfrac{1}{h_i} + \dfrac{x_{1,rr}}{2k_{ins,r}}} A_{ins,r}(T_{bs} - T_{ins,r}) \\ &\quad - k_{win} 0.33 A_1 (T_{bs} - T_{out}) \end{aligned} \tag{3.22}$$

To account for windows, equations (3.10) to (3.12) are modified as equations (3.23) to (3.25).

$$\rho_{ins} C_{p\_ins} V_{ins} \frac{dT_{ins}}{dt} = \frac{1}{\dfrac{1}{h_i} + \dfrac{x_1}{2k_{ins}}} 0.7 A_1 \left(T_{bs} - T_{ins}\right) - \ldots$$

$$\frac{2k_{ins} k_{conc}}{x_1 k_{conc} + x_2 k_{ins}} \left(A_2 - 0.33 A_1\right)\left(T_{ins} - T_{conc}\right) \qquad (3.23)$$

$$\rho_{conc} C_{p\_conc} V_{conc} \frac{dT_{conc}}{dt} = \frac{2k_{ins} k_{conc}}{x_1 k_{conc} + x_2 k_{ins}} \left(A_2 - 0.33 A_1\right)\left(T_{ins} - T_{conc}\right) - \ldots$$

$$\frac{2k_{conc} k_b}{x_2 k_b + x_3 k_{conc}} \left(A_3 - 0.33 A_1\right)\left(T_{conc} - T_b\right) \qquad (3.24)$$

$$\rho_b C_{p\_b} V_b \frac{dT_b}{dt} = \frac{2k_b k_{conc}}{x_3 k_{conc} + x_2 k_b} \left(A_3 - 0.33 A_1\right)\left(T_{conc} - T_b\right) - \ldots$$

$$\frac{1}{\dfrac{1}{h_o} + \dfrac{x_3}{2k_b}} \left(A_4 - 0.33 A_1\right)\left(T_b - T_{out}\right) \qquad (3.25)$$

The building energy system model developed so far has a roof with zero thermal mass. In the governing equations for the building energy system model (equation (3.23) to equation (3.25)), the effect of the mass of the roof on the building space temperature is neglected. However, this is not the case in practice. In order to address such a gap, an imperfectly insulated roof on one side of the thermodynamic control volume under study (i.e., the building space volume) with a mass of $\rho_{conc} V_{roof\_conc}$ is embedded into the building energy system model. The roof is a single-layer building construction element made of concrete with the same thermophysical properties as used for equation (3.23) to equation (3.25), and the outer side is exposed to the outdoor environment. The energy transfer equation for the roof is given as equation (3.26).

$$\rho_{conc} C_{p\_conc} V_{roof\_conc} \frac{dT_{roof}}{dt} = \frac{1}{\dfrac{1}{h_i} + \dfrac{x_{roof}}{2k_{conc}}} A_{roof} \left(T_{bs} - T_{roof}\right) - \ldots$$

$$\frac{1}{\dfrac{1}{h_o} + \dfrac{x_{roof}}{2k_{conc}}} A_{roof} \left(T_{roof} - T_{out}\right) \qquad (3.26)$$

In the case of the multi-storey modelling of buildings, where the roof of a particular building space under study is not exposed to the outside environment but to a building space of a different zone, then T_{out} in equation (3.26) is replaced by the surface temperature of the roof (other side) to which the building space air is exposed.

3.4 DYNAMIC RESPONSE FOR THE DEVELOPED BUILDING ENERGY SYSTEM MODEL

Building energy system models developed in the form of state-space (equation 3.20 and 3.21) are simulated in MATLAB/Simulink with step input excitations. A step function, $u(t)$, is represented as shown in Figure 3.12.

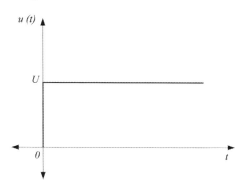

Figure 3.12 Step excitation/input to the building energy system model.

A step input function is defined as equation (3.27).

$$u(t) = \begin{pmatrix} 0 & \forall t < 0 \\ U & \forall t \geq 0 \end{pmatrix} \tag{3.27}$$

The numerical values of the parameters driving the building energy system model are taken from ASHRAE (2018) and Harish & Kumar (2016) and are presented in Table 3.2. A, B, C, and D matrices of the building energy state-space model are given as equations (3.28) to (3.31).

$$[A] = \begin{bmatrix} -6.65 \times 10^{-3} & 6.32 \times 10^{-3} & 0 & 0 \\ 2.99 \times 10^{-5} & -3.23 \times 10^{-5} & 2.43 \times 10^{-6} & 0 \\ 0 & 9.15 \times 10^{-5} & -1.9 \times 10^{-4} & 9.89 \times 10^{-5} \\ 0 & 0 & 2.98 \times 10^{-6} & -5.35 \times 10^{-5} \end{bmatrix} \tag{3.28}$$

$$[B] = \begin{bmatrix} 2.31 \times 10^{-5} & 0 & 3.33 \times 10^{-4} \\ 0 & 0 & 0 \\ 0 & 0 & 0 \\ 0 & 5.05 \times 10^{-5} & 0 \end{bmatrix} \tag{3.29}$$

$$[C] = \begin{bmatrix} 1 & 0 & 0 & 0 \end{bmatrix} \tag{3.30}$$

$$[D] = \begin{bmatrix} 0 \end{bmatrix} \tag{3.31}$$

Using equation (3.21), the system transfer function for the building energy system model is given as equation (3.32) to equation (3.34).

$$G_1(s) = \frac{T_{bs}}{\dot{Q}} = \frac{7.1 \times 10^{-18} + 4.049 \times 10^{-13} s + 6.378 \times 10^{-9} s^2 + 2.31 \times 10^{-5} s^3}{1.791 \times 10^{-16} + 7.091 \times 10^{-11} s + 1.666 \times 10^{-6} s^2 + 0.006929 s^3 + s^4} \quad (3.32)$$

$$G_2(s) = \frac{T_{bs}}{T_{out}} = \frac{7.67 \times 10^{-17}}{1.791 \times 10^{-16} + 7.091 \times 10^{-11} s + 1.666 \times 10^{-6} s^2 + 0.006929 s^3 + s^4} \quad (3.33)$$

$$G_2(s) = \frac{T_{bs}}{T_{Vent}} = \frac{1.0189 \times 10^{-16} + 5.84 \times 10^{-12} s + 9.184 \times 10^{-8} s^2 + 3.33 \times 10^{-4} s^3}{1.791 \times 10^{-16} + 7.091 \times 10^{-11} s + 1.666 \times 10^{-6} s^2 + 0.006929 s^3 + s^4} \quad (3.34)$$

The step responses for the system transfer functions of equation (3.32) to equation (3.34) are given in Figure 3.13 to Figure 3.15.

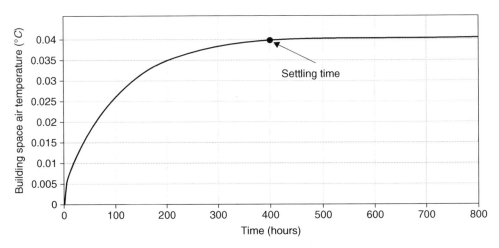

Figure 3.13 Response of the building energy system model for step excitation of heat rate of an HVAC system.

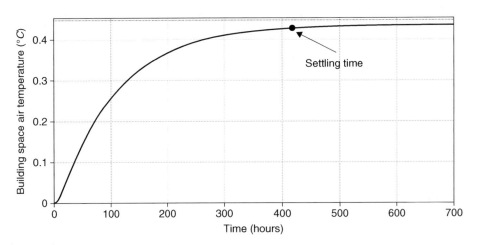

Figure 3.14 Response of the building energy system model for step excitation of outdoor air temperature.

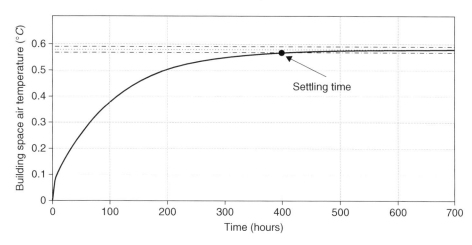

Figure 3.15 Response of the building energy system model for step excitation of air temperature ventilated into the building space.

Building energy system model performance analysis is carried out by specifying the response characteristics of settling time and rise time, etc. The performance characteristics for the building energy system is given in Table 3.3.

Table 3.3 Performance characteristics of the system responses

Characteristics	$G_1(s)$	$G_2(s)$	$G_3(s)$
Rise time	228.37 hours	230.87 hours	228.37 hours
Settling time	401.03 hours	417.47 hours	401.03 hours
Settling minimum	0.0364 °C	0.3937 °C	0.5256 °C
Settling maximum	0.0401 °C	0.4360 °C	0.5809 °C
Peak value	0.0411 °C	0.4360 °C	0.5809 °C
Peak time	768.17 hours	768.17 hours	768.17 hours

The building energy system model under study is excited with step excitations of outdoor air temperature, (T_{out}), ventilated air temperature, (T_{vent}), and HVAC plant heat rate, $(\dot{Q})$, and the level of HVAC heat rate or heating power is analyzed for different scenarios by varying the excitation values as shown in Table 3.4.

Table 3.4 Performance analysis by varying excitation values for different scenarios

Scenario	T_{out} (°C)	T_{vent} (°C)	T_{BS} (°C)	$\dot{Q}$(W/s)
I	0	0	20	505
II	0	17	9.7	0
III	0	17	20	255

In order to maintain the building space air temperature, (T_{BS}), at 20 °C, 505 W of HVAC power is required under zero outdoor and ventilated air temperature values. When there is no HVAC power, the building space temperature is 9.7 °C with a ventilated air temperature of 17 °C and zero outdoor temperature. Under zero outdoor temperature conditions, 255 W of HVAC power is required to maintain the building space air temperature and ventilated air temperature at 20 °C and 17 °C, respectively. The HVAC plant heat rate, ($\dot{Q}$), is calculated using energy balance equations as given in equations (3.35) and (3.36).

$$Q = \sum (UA_2) \times (T_{bs} - T_{out}) + V_{bs} \cdot \rho_{bs} \cdot C_{p\_bs} \cdot (T_{bs} - T_{vent}) W \tag{3.35}$$

and

$$\sum (UA_2) = \frac{A_2}{\frac{1}{h_i} + \frac{x_1}{k_{ins}} + \frac{x_2}{k_{conc}} + \frac{x_1}{k_b} + \frac{1}{h_o}} \tag{3.36}$$

Assuming that no share of the building space load from the lighting system is directly absorbed in the return airstream without becoming the building space's cooling load, the instantaneous building space load due to the lighting system is calculated as equation (3.37).

$$Q_{Light} = 8 \times 8 \times 1 \times 0.85 = 54.4 W \tag{3.37}$$

where

q_{Light}	=	$8W$
N	=	8
F_{use}	=	1
F_{SA}	=	0.85

The controllable and uncontrollable parameters driving the building energy system (BES) model are shown in Figure 3.16.

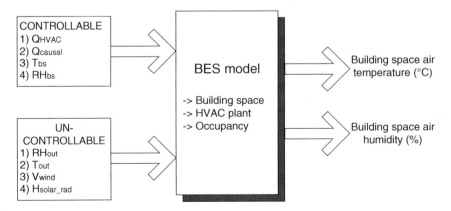

Figure 3.16 Building energy system model structure with inputs, operational parameters, and outputs.

In spaces with movable louvres and blinds, solar radiation ($H_{solar\ rad}$) is regarded as a partially controllable energy input to the building energy system model. However, in the present study, the influence of $H_{solar\ rad}$ is significant and the building energy system model under study is composed of windows that can be opened or closed, enabling full or no entrance of solar radiation through the glazing area. The uncontrollable parameters driving the building energy system model under study are shown in Figure 3.17 to Figure 3.20.

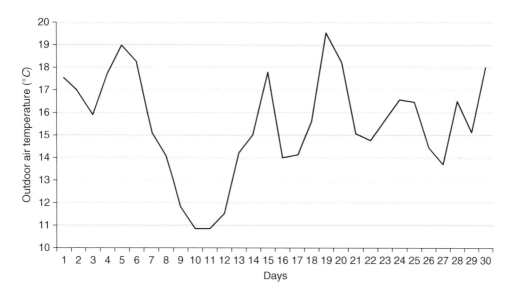

Figure 3.17 Outdoor air temperature (°C).

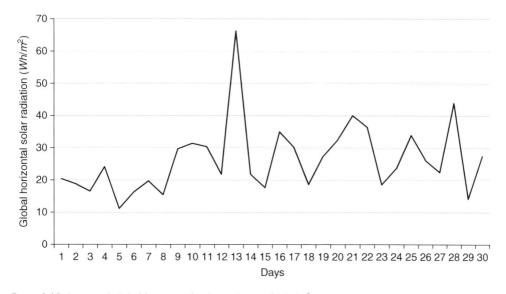

Figure 3.18 Averaged global horizontal solar radiation (Wh /m²) [176, 177].

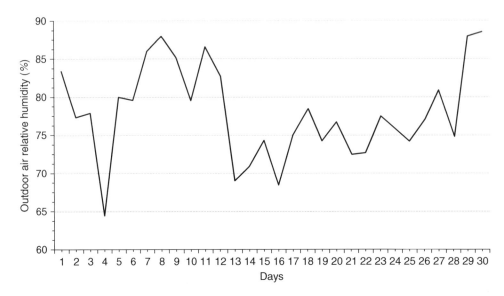

Figure 3.19 Outdoor relative humidity (%).

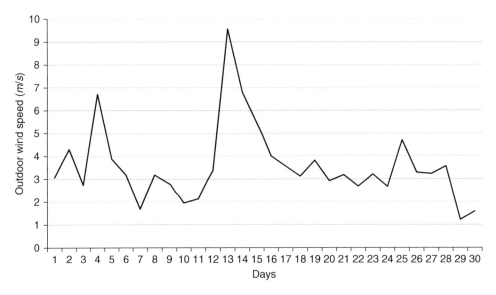

Figure 3.20 Outdoor wind speed (m/s) [176, 177].

A PID controller in an HVAC system is used to control the flow rate of the conditioned air into the building space and also to track the desired building space air temperature. In the present study, the thermostat's set point temperature has been set to 16.5 °C. This is a manual setting defined by occupants. The heat emitted to the building space by the HVAC system designed to supply a maximum flow rate of 0.15 *kg/s* at a temperature of 17.8 °C is shown in Figure 3.21.

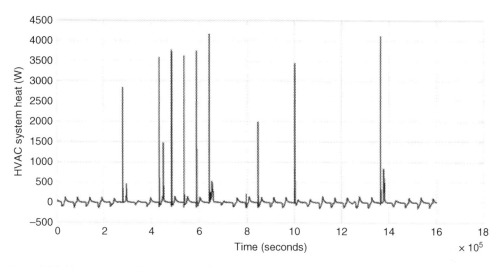

Figure 3.21 Heat emission from the HVAC plant to the building space (W).

The heat emitted to the building space from the HVAC plant varies between negative and positive values. Positive values indicate that the HVAC plant is heating the building space (heating mode) and negative values indicate cooling mode. This is varied in accordance with the outdoor conditions and the pre-defined thermostat set point temperature. A building energy system model is simulated for a one-month period and the variations in building space air temperature are shown in Figure 3.22.

The building space air temperature value regulates about a large difference. This is a peculiar property of a PID controller driven HVAC system. AI-tuned PID controllers are used nowadays for better regulation of building space air temperature. The current study focuses on the development of control strategies for optimal energy control and occupants' comfort management in buildings. These strategies can be applied to any building irrespective of its functionality and the type of HVAC control being used. The building space relative humidity for a one-month simulation is shown in Figure 3.23.

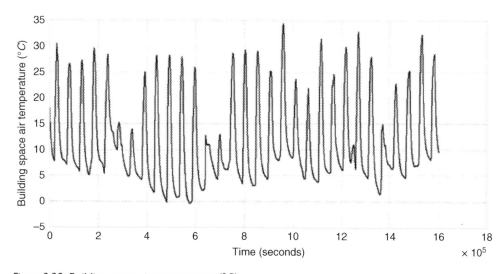

Figure 3.22 Building space air temperature (°C).

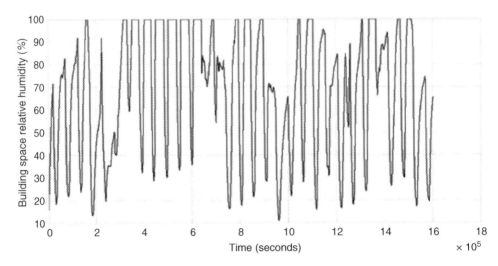

Figure 3.23 Building space relative humidity (%).

REFERENCES

Harish, V. S. K. V., & Kumar, A. (2014, January). Techniques used to construct an energy model for attaining energy efficiency in building: A review. In *Proceedings of The 2014 International Conference on Control, Instrumentation, Energy and Communication (CIEC)* (pp. 366–370). IEEE, doi.org/10.1109/CIEC.2014.6959111.

Harish, V. S. K. V., & Kumar, A. (2016). A review on modeling and simulation of building energy systems. *Renewable and Sustainable Energy Reviews*, 56, 1272–1292, doi.org/10.1016/j.rser.2015.12.040.

Harish, V. S. K. V., & Kumar, A. (2016). Reduced order modeling and parameter identification of a building energy system model through an optimization routine. *Applied Energy*, 162, 1010–1023, doi.org/10.1016/j.apenergy.2015.10.137.

Harish, V. S. K. V., & Kumar, A. (2016, January). Modeling and simulation of a simple building energy system. In *2016 International Conference on Microelectronics, Computing and Communications (MicroCom)* (pp. 1–6). IEEE, doi.org/10.1109/MicroCom.2016.7522473.

Harish, V. S. K. V., & Kumar, A. (2019, October). Stability analysis of reduced order building energy models for optimal energy control. In *2019 2nd International Conference on Power Energy, Environment and Intelligent Control (PEEIC)* (pp. 327–331). IEEE, doi.org/10.1109/PEEIC47157.2019.8976699.

Harish, V. S. K. V., Sant, A. V., & Kumar, A. (2021). Determining the performance characteristics of a white-box building energy system model and evaluating the energy consumption. In *Advances in Clean Energy Technologies* (pp. 605–615). Springer, Singapore, doi.org/10.1007/978-981-16-0235-1_46.

Underwood, C., & Yik, F. (2008). *Modelling Methods for Energy in Buildings*. John Wiley & Sons, doi.org/10.1002/9780470758533.

Underwood, D. (2015). Personal communication.

Chapter 4

Review of power converters

Amit Vilas Sant, Meet Patel, and V.S.K.V. Harish

CONTENTS

4.1 POWER CONVERTERS

Power electronics can be broadly defined as the branch of science and technology that deals with the study and application of power semiconductor switches and power converters for power modulation as per the requirements while ensuring efficient power transfer and precise load operation. Power electronic switches are the basic building blocks of power converters. Power converters can facilitate the conversion of ac power to dc power and vice versa. In addition, the levels of dc voltage and current can be modulated with dc–dc converters. For an ac system, an ac–ac converter can work to control the phase, frequency, amplitude, or all three quantities at the output. With the energy-efficient control of the flow of power and static operation, power converters are widely employed in the domestic, commercial, industrial, transportation, and utility sectors. In the domestic and commercial sectors, power converters find application in LED lighting, fan regulators, electronic ballasts, refrigeration units, air-conditioners, battery chargers, laptops, desktop computers, elevators, uninterruptible power supplies, etc. In industrial plants, adjustable speed drives are extensively used. Adjustable speed drives comprise an ac–dc converter feeding a dc–ac power converter. In the transportation sector, power converters are used in propulsion systems as well as battery chargers. Power converters find applications even in electric aircraft, electric ships, and satellites. Over the past two to three decades, the utility sector has seen tremendous developments in the form of grid integration of renewable energy sources, distributed generation, high voltage dc transmission systems, custom power, FACTS devices, etc. All of the areas mentioned involve the use of one or more power converters.

Solar and wind energy are employed for electric power generation. More and more countries are implementing legislative measures to facilitate and encourage renewable energy system-based electric power generation. Wind energy conversion systems harvest the kinetic energy of the wind, converting it into rotational energy and finally into electrical energy. Wind turbines are responsible for the conversion of the kinetic energy of the wind into mechanical energy. Similarly, electric generators convert this mechanical energy into electrical energy. Moreover, only a small operating range is available for electric power generation. In solar systems, photovoltaic panels are employed to convert solar energy into electrical energy. The output of the photovoltaic panel is dc, whereas the grid, as well as the majority of the loads, requires ac supply. Electric power generation from renewable energy sources is highly dependent on environmental conditions.

The integration of these renewable energy sources with utility networks requires strict adherence to the established grid codes. The electric power generated by solar and wind energy systems needs to be regulated and modulated before being integrated with the grid. Power converters play a significant role in the grid integration of renewable energy sources. It would not be farfetched to say that the grid integration of renewable energy sources cannot be possible without power converters. Power semiconductor switches are the basic building block of any power converter. This chapter presents a review of power semiconductor switches and power converters. A detailed classification of power converters and their applications is also presented.

4.2 IDEAL AND NON-IDEAL SWITCHES

In mechanical switches, such as toggle switches and push-button switches, by applying force, the switching state can be changed. These switches have moving parts and take at least hundreds of milliseconds to change state. The life of the switch is given in terms of number of switching operations. Arcing is also observed while changing switching states. Moreover, human interference is needed to change the switching state. Mechanical relays can overcome this by incorporating automated operation. The relay has normally closed (NC) and normally open (NO) contacts. By energizing the control coil, NO contacts can be closed and NC contacts can be opened. Thus, the application of voltage across the control coil results in control over the operating state of the relay. Though mechanical relays offer a low-cost automated solution, they suffer from drawbacks in that the time needed to switch from one state to another is around 200 milliseconds, and moving parts are involved.

An ideal switch should have no moving parts and no power losses and switch from one state to another instantaneously. Figures 4.1 and 4.2 show the ideal switch connected in a dc circuit and the current and voltage waveforms recorded across the load and switch.

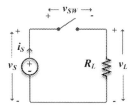

Figure 4.1 Ideal switch connected in a dc circuit.

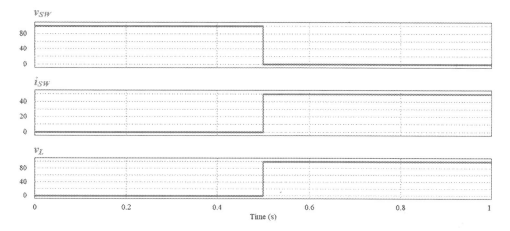

Figure 4.2 Switch voltage, switch current and load voltage.

By Kirchoff's voltage law,

$$v_S = v_{SW} + v_L \tag{4.1}$$

Where v_s is the supply voltage, v_{sw} is the voltage across the switch, and v_L is the load voltage.

The ideal switch acts as a short circuit (i.e., the resistance of the switch is zero) when it is ON. With a switch acting as a short-circuited path, v_{SW} is zero. Hence, as per equation 4.1, v_S is applied at the load terminals. In this circuit, the current supplied by the source and consumed by the load, as well as current flowing through the switch, are equal and can be determined by Ohm's law.

On the other hand, the ideal switch is analogous to an open circuit and offers infinite resistance when it is OFF. In this situation, the load voltage is reduced to zero (i.e., $v_L = 0$) as no current flows through the circuit as the switch offers infinite resistance. The supply voltage appears across the switch (i.e., $v_{SW} = v_S$).

Power semiconductor switches are mainly silicon-based switches with no moving parts. Generally, these switches have two power terminals and a control terminal. The signal applied at the control terminal is the activation signal. Upon application of the activation signal, the switch starts to conduct and acts as a closed switch if it is forward-biased. In the absence of a closed activation signal, the switch may or may not stop conducting, depending on its construction. However, the switch stops conducting when it is reverse-biased. Moreover, an activation signal cannot turn ON the reverse-biased switch. Though power semiconductor switches are static switches, they are far from ideal switches. When this switch is ON, it does not offer zero resistance; rather, the ohmic property of the silicon layers in the switch comes into the picture to offer a small resistance. This resistance is known as ON-state resistance, and the voltage drop occurring across the switch on account of this ON-state resistance is known as ON-state voltage drop. Also, when this switch is OFF and forward-biased, ideally it should offer infinite resistance and not allow any current to flow through the switch. However, current with very small magnitude flows through the switch. This current is known as forward leakage current. As the forward leakage current flows through the switch during the ON state, it contributes to I^2R losses, generally known as conduction losses. The switch offers very high resistance when it is reverse-biased. A small current flows through the switch, albeit in the reverse direction as compared to the forward leakage current. This current is termed reverse leakage current.

Figures 4.3 and 4.4 show a circuit diagram with a power semiconductor switch and the waveforms for switch voltage and current. Unlike ideal switches, the power semiconductor switches take a finite time to turn ON or OFF. The turn-ON period is the duration from the instant of the application of the activation signal to the instant at which the voltage across the switch falls to the ON-state value. Similarly, the turn-OFF period is the duration from the instant of the removal of the gate pulse or reverse biasing to the instant when the current through the switch drops to the leakage value. During these two switching periods, the product of current and voltage has non-zero values. The sum of the non-zero power loss during the switching periods is termed the switching losses. Conduction losses are dependent on the ON-state voltage drop, v_{dON}, or ON-state resistance, r_{dON}, and the current flowing through the switch, i_{sw}. Switching losses depend on the energy required to switch the switch ON and OFF, respectively. Moreover, switching losses are also dependent on switching frequency. The switching loss, P_{SWloss}, and conduction loss, P_{Closs}, can be given as:

$$P_{Closs} = i_{SW} v_{dON} = i_{SW}{}^2 r_{dON} \tag{4.2}$$

$$P_{SWloss} = f_{SW} \left(v_{dON} \right) = i_{SW}{}^2 r_{dON} \tag{4.3}$$

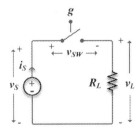

Figure 4.3 Ideal switch connected in a dc circuit.

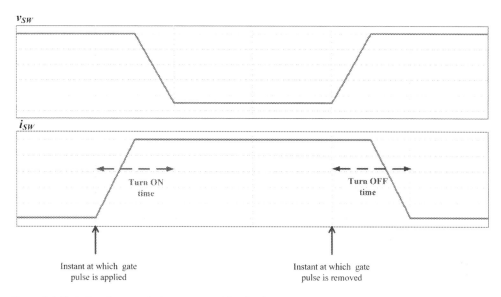

Figure 4.4 Switch voltage, switch current and load voltage.

4.3 CLASSIFICATION OF POWER SEMICONDUCTOR SWITCHES

Power semiconductor switches involve a fast switching action, static operation, and automated operation based on a control signal. Different power semiconductor switches are commercially available. Switches are selected based on the application and power converter. Each switch has different features and characteristics; however, some common points are:

(i)   Power semiconductor switches are designed to conduct only in a forward-biased state.
(ii) Power semiconductor switches are designed to block the voltage in a reverse-biased state.
(iii) Power semiconductor switches have two power terminals through which power is transferred from the supply to the load.
(iv) In most cases, power semiconductor switches have a control terminal. The application of a voltage signal having an amplitude above the threshold value forces the forward-biased switch to conduct.
(v)   In its ON-state, a power semiconductor switch acts as a closed switch and ideally offers zero resistance. In practice, a voltage drop across the switch is observed while it is conducting. This voltage drop is of very small magnitude; however, it is significant and cannot be neglected.
(vi) In its OFF-state, the power semiconductor switch acts as a closed switch and blocks the voltage. Ideally, the switch offers zero resistance; however, in practice, forward leakage current (in the range of a few milliamperes) flows through the forward-biased power semiconductor switch and reverse leakage current (in the range of a few milliamperes) flows through the reverse-biased switch.

Details about some of the commonly used power semiconductor switches are given below.

1. Power diode
    a. Symbol

Anode

Cathode

    b. Power terminals: 02 (anode and cathode)
    c. Control terminals: None
    d. ON-state and OFF-state: A power diode is considered to be forward-biased when the anode has higher potential than the cathode. As soon as the forward breakover voltage is exceeded, the forward-biased power diode enters conduction mode. The power diode can conduct in a reverse-biased state when the reverse breakover voltage is exceeded. However, this would result in permanent damage to the diode due to excessive heating. Conversely, the diode is reverse-biased when the cathode is at a higher potential than the anode, and it blocks the voltage until the reverse breakover voltage is exceeded.
    e. Control: A power diode is a unidirectional voltage-controlled device. There is no control over the instant of turn ON or turn OFF. When a power diode is forward-biased and the forward breakover voltage is exceeded, it conducts. When the diode is reverse-biased, it turns OFF.
    f. Application in power converters: Uncontrolled rectifiers, semi-controlled rectifiers, choppers, switched-mode power supplies, inverters.
2. Silicon controlled rectifier (SCR)
    a. Symbol

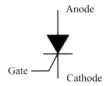

Anode

Gate

Cathode

    b. Power terminals: 02 (anode and cathode)
    c. Control terminals: 01 (gate)
    d. ON-state and OFF-state: Just as with a power diode, an SCR is considered forward-biased at the moment its anode is relatively at higher potential. Upon application of gate current that exceeds the threshold value, the forward-biased SCR starts conducting. Alternately, the SCR may be turned ON by exceeding the rated forward breakover voltage. In any application, the SCR is selected such that the rated forward breakover voltage is not exceeded. By applying the gate pulse to the gate terminal, the forward breakover voltage can be decreased. The higher the gate current, the lower the breakdown voltage. With the application of the maximum permissible gate current, the forward-biased SCR can be instantly turned ON. Similar to the power diode, the SCR can also be turned ON under

a reverse-biased state, but it is undesirable as the SCR would be permanently damaged due to excessive heating. As the rated forward breakover voltage is not to be exceeded, the forward-biased SCR blocks the voltage until the application of the gate pulse. The reverse-biased SCR blocks the voltage until the reverse breakover voltage is exceeded.

e. Control: An SCR is a unidirectional current controlled device with control only over the instant of turn ON during the forward-biased state. The instant of turn ON can be controlled with the gate pulse. However, once the SCR starts conducting, the control is lost. When the anode to cathode current is reduced below the holding current, the SCR is turned OFF. In short, the holding current is the minimum value of the anode to cathode current that must flow through an SCR operating in a forward conduction state for the SCR to continue operating in this state. The SCR operating in forward conduction mode can be turned OFF with natural commutation through the reversal of polarity of the supply voltage or through forced commutation (i.e., the use of passive elements, diodes, or SCR to force the anode to cathode current to reduce below the holding current). Further, reverse voltage needs to be applied across the SCR to remove the free charge carriers and regain the forward blocking capability. SCRs with a current rating as high as 1.7 kA and a voltage rating as high as 3.6 kV are available in the market. The switching frequency of a converter grade SCR is around 600Hz, and for an inverter grade SCR it is around 3kHz.

f. Application in power converters: Phase-controlled rectifiers, choppers

3. Diode for alternating current (DIAC)

a. Symbol

Terminal A$_1$

Terminal A$_2$

b. Power terminals:02 (terminal A$_1$ and terminal A$_2$)

c. Control terminals: None

d. Bidirectional device

e. ON-state and OFF-state: A DIAC is a bidirectional device. When the voltage across the DIAC exceeds the threshold value, it starts conducting. Current flows from terminal A$_1$ to A$_2$ or A$_2$ to A$_1$ depending on which terminal is at higher polarity. Some of the typical values of the threshold voltage are 16V or 32V.

f. Control: A DIAC is similar to an antiparallel connection of two diodes. The only difference is in terms of the higher value of threshold voltage. A DIAC is a voltage controlled bidirectional device with no control over the instant of turn ON or turn OFF. When the voltage exceeds the threshold value, the DIAC starts to conduct, and when it is reverse-biased it stops conducting.

g. Application in power converters: A DIAC is employed in the firing circuit for the TRIAC-based ac voltage controller employed in fan regulators.

4. Triode for alternating current (TRIAC)
   a. Symbol

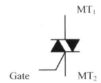

   b. Power terminals: 02 (MT_1 terminal and MT_2 terminal)
   c. Control terminals: 01 (gate)
   d. ON-state: In contrast to SCRs, TRIACs allow current to flow in both directions. A positive or negative gate pulse can be used to trigger a TRIAC; however, only a positive gate pulse is required for an SCR. It should be noted that the gate pulse is applied between MT_1 and the gate terminal. Even if the gate current interrupts, a TRIAC continues to conduct until the anode to cathode current is decreased below the holding current. A TRIAC remains in an OFF condition irrespective of the polarity of the terminals unless a gate pulse is applied.
   e. Control: A TRIAC is a current-controlled bidirectional device with control over the instant of turn ON. TRIACs with a voltage rating of up to 1.2kV are commercially available. A TRIAC operates in four different operating modes depending on the polarity of MT_1 with respect to MT_2 and that of the gate current.
5. Power bipolar junction transistor (BJT)
   a. Symbol

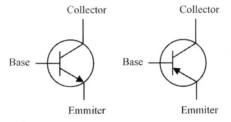

   b. Power terminals: 02 (collector and emitter)
   c. Control terminals: 01 (base)
   d. ON-state: A power BJT can have three operating states: saturation, cut-off, and active region. When the power BJT is operated in saturation and cut-off region only, it is said to be operating as a switch. A power BJT operating in a cut-off region acts as an open switch. Conversely, a power BJT operating in a saturation region acts as a closed switch. In an NPN configuration, when the collector is at a higher potential than the emitter, the device is said to be forward-biased. In the absence of base current, the forward-biased device is in the cut-off region and does not conduct. The entire voltage is dropped across the collector and emitter terminals. When the threshold base current is applied, the device enters the saturation region and the device starts conducting. A small fraction of applied voltage is dropped across the collector emitter terminals. In this configuration, with the emitter at a higher potential than the collector, the device is said to be reverse-biased.

e. Control: A power BJT is a unidirectional current controlled device. Unlike an SCR, in a power BJT there is control over the instant of turn ON as well as the instant of turn OFF. The instant of turn ON can be controlled with the base current. With the interruption of the base current, the power BJT enters the cut-off region, i.e., OFF state. However, the low current gain results in the need for higher gate currents, which has led to its application in power converters being severely limited. Power BJTs in the range of 25A and 700V are easily available in the market.

f. Application in power converters: Power BJTs can be employed in inverters. However, the low current gain of less than five necessitates a higher gate current. This leads to higher gating losses, and the power required by the gating circuitry increases with the amplitude of load current.

6. Power metal–oxide–semiconductor field-effect transistor (MOSFET)

a. Symbol:

b. Power terminals: 02 (drain and source)

c. Control terminals: 01 (gate)

d. ON-state and OFF-state: When the drain terminal is at a higher potential compared to the source terminal, the device is said to be forward-biased. However, due to the large ohmic resistance offered by the device, only a small leakage current flows and the device operates in an OFF state. When a positive voltage is applied between the gate and source terminal, a virtual n-channel is created. This virtual channel offers an alternate path with low resistance for the drain-to-source current. The higher the amplitude of the gate voltage, the higher the width of the virtual n-channel and the lower the ohmic resistance being offered by the device. The device acts as an open switch and allows for the flow of current from drain to source. The presence of an antiparallel diode inherent to the device is to be noted. This diode starts conducting upon the application of a reverse voltage across it.

e. Control: A power MOSFET is a unidirectional voltage-controlled device with an inherent antiparallel diode. As in the case of the power BJT, a power MOSFET has control over the instant of switching ON and switching OFF. When the power MOSFET is forward-biased, the application of a gate pulse results in the device operating in a conduction state, and when the gate pulse is zero, the device reverts to an OFF state. The SiO_2 layer in the structure of the device behaves as a dielectric medium and allows a gate current in the nanoamperes range. Hence, the gating losses are negligible. During the conduction state, the higher ON-state resistance results in increased conduction losses. For IRF740, the ON-state resistance is around 0.55Ω. This has resulted in power MOSFETs finding applications in the low and medium power range only. The switching frequency of the power MOSFET can be in the hundreds of kilohertz.

     f.   Application in power converters: Power MOSFETs are employed in inverters with lower power ratings. They are extensively used in switched-mode power supplies where a higher switching frequency is required, and the power rating is usually low. They are also used in low-power pulse-width modulators employed as front-end converters.

7. Insulated-gate bipolar transistor (IGBT)

    a.   Symbol

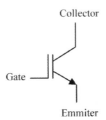

    b.   Power terminals: 02 (collector and emitter)

    c.   Control terminals: 01 (gate)

    d.   ON-state and OFF-state: From a gating perspective, IGBTs act similarly to power MOSFETs, whereas from the perspective of the flow of power, IGBTs are closer to power BJTs. Just as in the case of power BJTs, when the collector has a higher potential as compared to the emitter, the device is said to be forward-biased. With the emitter terminal and collector terminal at a lower potential, the IGBT is forward-biased. The IGBT does not conduct until the application of a gate pulse. Once a gate pulse is applied, the IGBT turns ON and continues to conduct until the gate pulse is removed. Only a forward-biased IGBT can conduct. When the IGBT is conducting, a small fraction of voltage is dropped across the collected emitter terminal. Rather than large ON-state conduction loss, as in a power MOSFET, an IGBT offers a smaller ON-state conduction loss similar to that observed in the case of a power BJT. Commercially available IGBTs have an antiparallel diode. When the IGBT is reverse-biased, the antiparallel diode starts to conduct.

    e.   Control: An IGBT can be triggered into conduction mode by applying a voltage pulse at its gate terminal. It is a unidirectional device. An IGBT behaves similarly to a MOSFET with respect to the gating requirements, but the issue of higher conduction loss in the MOSFET is not observed. For power transfer during conduction mode, its behaviour is similar to that of a power BJT. Thus, it can be said that IGBTs offer the benefits of power MOSFETs in terms of low gating power requirements and of power BJTs in terms of lower conduction losses. Moreover, similar to power BJTs and power MOSFETs, IGBTs allow for control over the instant of turn ON and turn OFF. For a forward-biased IGBT, when a pulse is applied, the IGBT conducts and acts as a closed switch until the gate pulse is not removed. IGBTs with current ratings as high as 3600A and voltage ratings as high as 6.5kV are reported.

    f.   Application in power converters: IGBTs are extensively used in inverters, whether two-level or multilevel. Moreover, in switched-mode power supplies for higher power ratings and in pulse-width modulated rectifiers, IGBTs are widely preferred. In matrix converters, bidirectional switches are required. An IGBT, being a unidirectional switch, cannot be directly used. Two IGBTs connected in series in a common emitter configuration are employed in matrix converters. Switching frequencies as high as 50kHz are reported for IGBTs.

8. Gate turn-off thyristor (GTO)
   a. Symbol

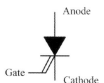

   b. Power terminals: 02 (anode and cathode)
   c. Control terminals: 01 (gate)
   d. ON-state and OFF-state: AGTO is similar to an SCR and can be turned ON upon the application of a gate pulse provided that the device is forward-biased. A GTO can be made to block the voltage with negative voltage at the gate terminal provided that a high enough negative gate current flows. It is not designed to conduct in reverse-biased mode.
   e. Control: A GTO is a unidirectional current-controlled device. With the application of a gate pulse, the conduction period can be controlled. Similar to an SCR, the instant of turn ON can be controlled with the gate pulse. Unlike an SCR, once a GTO starts conducting, the control is not lost. With a negative gate current, a GTO can be commutated. The required magnitude is often as high as one-third of the anode current being switched OFF.

9. Light-activated silicon-controlled rectifier (LASCR)
   a. Symbol

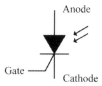

   b. Power terminals: 02 (anode and cathode)
   c. Control terminals: 01 (gate)
   d. ON-state and OFF-state: An LASCR functions like an SCR, with the only difference being that it is triggered by light directed at the gate terminal. For forward and reverse-bias conditions, the same rules that apply to an SCR hold true for an LASCR. When the LASCR is forward-biased, it is in an OFF state until the optical signal is not directed at the gate terminal. When the optical signal is applied at the gate of a forward-biased LASCR, it starts conducting and behaves as a closed switch. As in the case of an SCR, an LASCR continues to conduct even after the removed optical signal. An LASCR offers complete electrical isolation between the high voltage power circuit and the light triggering source.

4.4 CLASSIFICATION OF POWER CONVERTERS

Power converters modulate power and thereby control the flow of power. Based on the type of input and output supply for the power converter and type of modulation performed, power converters can be classified in detail as:

i. ac–dc power converters
  a. uncontrolled rectifiers
  b. phase-controlled rectifiers
  c. pulse-width modulated rectifiers
  d. multipulse rectifiers
ii. dc–dc power converters
  a. choppers (class A, B, C, D, E)
  b. switched-mode power supplies
     • non-isolated: buck, boost, buck-boost, ćuk, SEPIC converter
     • isolated (involves high-frequency ferrite transformer): flyback, push-pull, half-bridge, full-bridge, forward, single active bridge, dual active bridge converter
iii. dc–ac power converters
  a. two-level inverter
  b. multilevel inverter (neutral point clamped/diode clamped, flying capacitor, cascaded)
  c. current source inverter/voltage source inverter
iv. ac–ac power converters
  a. ac voltage controller (single phase/three phase, phase angle control/integral cycle control)
  b. cycloconverter
  c. matrix converter (direct, indirect,sparse)

4.4.1 ac–dc converters

ac–dc conversion is implemented by rectifiers, which can be further classified as:

(i) uncontrolled rectifiers (power diodes are used as power electronic switches)
(ii) phase-controlled rectifiers (silicon-controlled rectifiers (SCRs) are used as power electronic switches)
(iii) pulse-width modulated rectifiers (MOSFET/IGBTs are used as power electronic switches)

4.4.1.1 Uncontrolled rectifier

Figure 4.5 shows the power structure of single-phase and three-phase uncontrolled rectifiers. Uncontrolled rectifiers are made up of power diodes. In power diodes, there is no control over the instant of turn ON and turn OFF. When the diode is forward-biased, it starts conducting. Figure 4.6 shows the instantaneous voltage measured across the output terminals of single-phase and three-phase diode bridge rectifiers feeding a resistive load. From this figure, it is clear that the output of the diode bridge rectifier is not pure dc but pulsating dc. In a single-phase half-wave rectifier, only the positive half cycle of the supply appears at the output, whereas in a single-phase full-wave rectifier, the absolute value of the supply voltage appears at the output. In a three-phase rectifier, the maximum absolute value of the three-phase supply voltage appears across the output terminals. By determining the area under the output voltage waveform for one cycle of supply voltage, the average output voltage can be known.

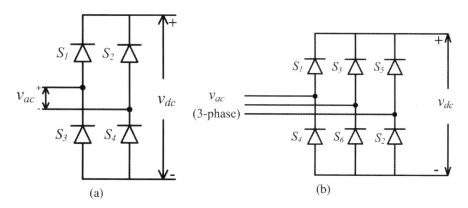

Figure 4.5 (a) single-phase diode bridge rectifier, and (b) three-phase diode bridge rectifier.

Equations (4.4) and (4.5), respectively, show the average output voltages for the two rectifiers, shown in Figure 4.6 (a) and (b). In these two equations, the average output voltage and maximum value of supply voltage are denoted by V_{out} and V_m, respectively. In these two equations, ω represents the supply frequency, and t denotes the time.

$$V_{out} = \frac{1}{2\pi} \int_{0}^{\pi} \left[V_m \sin(\omega t) \right] dt = \frac{V_m}{\pi} \tag{4.4}$$

$$V_{out} = \frac{1}{2\pi} \int_{0}^{\pi} \left[V_m \sin(\omega t) \right] dt = \frac{2V_m}{\pi} \tag{4.5}$$

From equations (4.4) and (4.5), it is clear that V_{out} is governed by V_m. As the conduction period of the diode is governed by the supply voltage and there is no control over the instant of turn ON and turn OFF, V_{out} cannot be controlled. This drawback can be overcome by replacing diodes with SCRs. The resulting power converter is an ac–dc converter known as a phase-controlled rectifier.

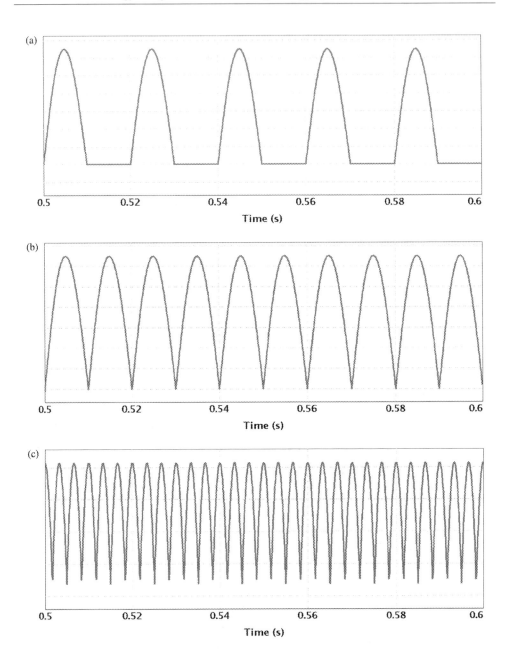

Figure 4.6   (a) output voltage of a single-phase half-wave diode bridge rectifier, (b) output voltage of a single-phase full-wave diode bridge rectifier, and (c) three-phase full-wave diode bridge rectifier.

4.4.1.2 Phase-controlled rectifier

The circuit for single-phase full-wave and three-phase full-wave-controlled rectifiers is shown in Figure 4.7. As mentioned earlier, phase-controlled rectifiers are made up of SCRs. In power diodes, there is no control over the instant of turn ON and turn OFF. A forward-biased SCR starts conducting when a gate pulse is applied and stops conducting when it is reverse-biased. The conduction period can range from 0 degrees to 180 degrees. The point of commutation for an SCR is fixed at 180 degrees from the start of the positive cycle of supply. In the case of diodes, the conduction starts when it is forward-biased. On the other hand, in an SCR, the instant of the start of conduction can be controlled; hence, the conduction period can also be controlled. As a consequence, the area under the curve can be controlled. By varying the firing angle, the area under the curve and the dc output voltage can be regulated. In this converter, the firing angle controls the output dc voltage. Figure 4.8 shows the waveforms for the output voltage of single-phase half-wave and full-wave phase-controlled rectifiers feeding a resistive load. As was also observed in the case of uncontrolled rectifiers, in phase-controlled rectifiers, the output dc voltage is pulsating. Undoubtedly, with the use of passive filters, the pulsating dc can be filtered to obtain a pure dc supply. With half-wave phase-controlled rectifiers, only a segment of the supply voltage when it has a positive polarity appears at the output. Whereas, with full-wave phase-controlled rectifiers, the segment of the supply voltage for both the polarities appears at the output. However, the power circuit ensures that for the segment of supply voltage for both positive and negative half cycles, the polarity of the output voltage is unchanged. The average value of the output voltage for half and full-wave rectifiers are respectively shown in equations (4.6) and (4.7), where V_{out} is the average value of output voltage, V_m is the maximum value of the supply voltage, α is the firing angle (i.e., the angle at which the SCR starts conducting), ω is the supply frequency, and t is the time.

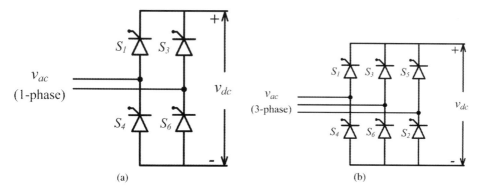

Figure 4.7 Phase controlled rectifier for (a) single-phase supply, and (b) three-phase supply.

$$V_{out} = \frac{1}{2\pi}\int_{\alpha}^{\pi}\left[V_m \sin(\omega t)\right]dt = \frac{V_m}{2\pi}(1+\cos\alpha) \tag{4.6}$$

$$V_{out} = \frac{1}{2\pi}\int_{\alpha}^{\pi}\left[V_m \sin(\omega t)\right]dt = \frac{V_m}{\pi}(1+\cos\alpha) \tag{4.7}$$

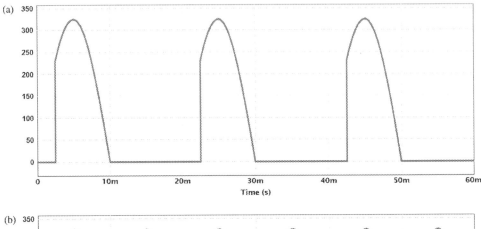

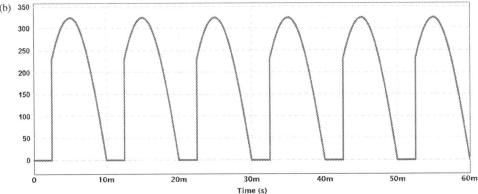

Figure 4.8   Output voltage of (a) single-phase half-wave phase-controlled rectifier, and (b) single-phase full-wave phase-controlled rectifier.

From equations (4.6) and (4.7), it is clear that the lower the value of α, the higher the value of $cos(\alpha)$, and consequently, the higher the value of V_{out}. As the conduction period of the SCR can be controlled by controlling α, the instant of turn ON, the area under the curve in the output voltage waveform, and the average value of output voltage can be controlled.

4.4.1.3 Pulse-width modulated rectifier

Figure 4.9 shows the currents being drawn by a three-phase uncontrolled rectifier feeding a resistive load. The current is non-sinusoidal, with a total harmonic distortion (THD) of around 28%. These non-sinusoidal currents comprise fundamental and harmonics components. The harmonic current flows through the grid and causes current distortions and consequently leads to voltage distortions. The resulting grid pollution has a detrimental effect on the efficiency of the distribution network as well as the loads connected to the network. To ensure that sinusoidal current is drawn by the system while facilitating ac–dc conversion and control over the magnitude of output dc voltage, pulse-width modulated rectifiers are employed.

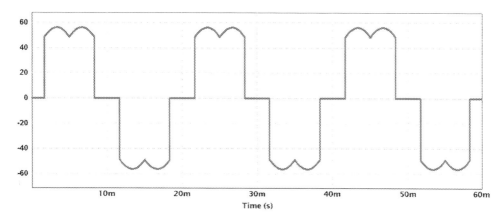

Figure 4.9 Current drawn from the grid by three-phase uncontrolled rectifier feeding a resistive load.

Figure 4.10 shows the power structure of a three-phase pulse-width modulated rectifier and output voltage. A filter capacitor needs to be connected at the output of this rectifier to filter out the ripples in the output dc voltage. Hence, the dc is almost constant with a very small ripple voltage. In order to achieve this, there should be control over the instant of turn ON and turn OFF of the power semiconductor switch. Hence, IGBTs and MOSFETs are employed. The inductor is connected at the input to filter out the current harmonics. In this power converter, pulse-width modulated gate pulses are applied to the IGBTs/MOSFETs; hence, the converter is termed a pulse-width modulated rectifier. The control system for the generation of gate pulses involves an outer dc link control loop and an inner ac current control loop. The output of an ac current control loop is the reference voltage, which is processed by the sinusoidal pulse width modulator for the generation of gate pulses. A phase-locked loop is also involved in the control algorithm for the estimation of the phase angle of the supply voltage. Besides regulating the dc link voltage at the reference value, the control action also ensures that the currents drawn from the supply are sinusoidal, with a THD of less than 5%, and in phase with the supply voltage. This would prevent the distortion of grid voltages as well as currents and have no adverse effect on the power quality of the grid.

To reduce the THD content in the currents being drawn from the grid by the rectifiers, a multipulse configuration can also be employed. These rectifiers involve multiple six-pulse rectifiers and transformers to achieve the ac–dc conversion while ensuring that the THD in the supply current is considerably reduced. 12, 18 and 24-pulse rectifiers are commonly reported. The higher the number of the pulse, p, the higher the order of the dominant harmonics in supply current $p \pm 1$, and the lower the THD content in the current.

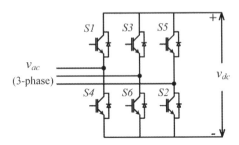

Figure 4.10 Pulse-width modulated rectifier.

4.4.2 dc–dc converters

dc–dc conversion is implemented by choppers and switched-mode power supplies. Generally, in dc–dc converters, the output voltage needs to be regulated. The two types of dc–dc converter are:

- choppers
- switched-mode power supplies (SMPS)

4.4.2.1 Choppers

Choppers employ SCRs. They are designed not to control the instantaneous value of an output voltage but its average value. Choppers were mainly designed to control the shaft speed of dc motors. Depending on the quadrant of operation, choppers can be classified as class A, B, C, D, and E. These chopper configurations are shown in Figure 4.11. When supplying the load, the output of a chopper varies between the supply voltage and zero. By controlling the ON and OFF times of the SCR, the average value of the output voltage is controlled. As the input is a dc supply, SCRs are forward-biased at all times. For turning OFF SCRs, additional commutation circuitry needs to be incorporated. Moreover, due to the use of SCRs, the use of higher switching frequencies is not possible. Choppers can only be employed for dc motors and similar high inertia loads that can withstand the variation in dc supply. As ac motors are largely preferred over dc motors, choppers find limited applications nowadays.

4.4.2.2 Switched-mode power supplies

SMPS are dc–dc converters that employ MOSFETs in low-power applications and IGBTs in high-power applications. Rather than the average value of output dc voltage, the instantaneous output voltage is controlled. With appropriate design and control, the output voltage ripple can be regulated within permissible limits (less than 5%). As IGBTs and MOSFETs are employed, commutation circuitry is not needed. Also, higher switching frequencies from 10–50 kHz can be easily employed. The higher the switching frequency, the lower the value of passive components and the more compact the converter design.

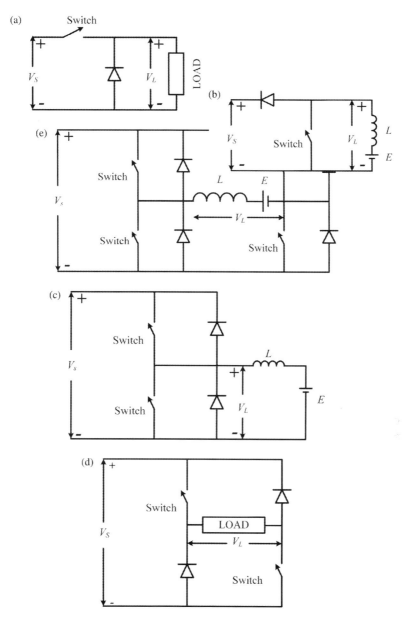

Figure 4.11 (a) Class A chopper, (b) Class B chopper, (c) Class C chopper, (d) Class D chopper, and (e) Class E chopper.

Non-Isolated Switched-Mode Power Supplies

SMPS can be classified into non-isolated and isolated configurations. The basic non-isolated converters, buck, boost, and buck-boost converters, are shown in Figure 4.12. These converters involve a power electronic switch, usually a MOSFET or IGBT, an inductor, a capacitor, and a diode. There is no involvement of any transformer. In a buck dc–dc converter, the input voltage is stepped down and made available at the output. The relationship between input and output dc voltage is given as:

$$V_{out} = V_{in}D \tag{4.8}$$

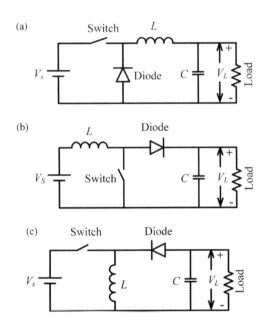

Figure 4.12 (a) Buck converter, (b) boost converter, and (c) buck-boost converter.

where V_{out} is the output voltage, V_{in} is the input voltage, and D is the duty cycle. D is the ratio of time period for which the switch was ON to the sum of the time periods for which the switch was ON and OFF. For boost converters and buck-boost converters, the relationship between V_{in} and V_{out} are given, respectively, as:

$$V_{out} = V_{in}/(1-D) \tag{4.9}$$

$$V_{out} = V_{in}D/(1-D) \tag{4.10}$$

In a boost converter, the variation in D from 0 to 1 results in V_{out} being changed from 0 to ∞. However, in practice, the maximum value of V_{out} is not more than two times V_{in}. In a buck-boost converter, for values of D less than 0.5, the converter behaves as a buck converter. Conversely, for values of D more than 0.5, the buck-boost converter behaves as a boost converter. With a value of D equal to 0.5, V_{out} for the buck-boost converter equals V_{in}. SEPIC and Ćukare the other non-isolated converter topologies.

Isolated Switched-Mode Power Supplies

In non-isolated SMPS, the input and output voltages are not isolated, and the boost factor is usually limited to two. Similarly, in buck operation, problems are faced at very low values of D. An isolated dc–dc converter includes a high-frequency ferrite core transformer for facilitating isolation between the input and output terminals, as well as to allow for the stepping up or stepping down of voltages without any limitations. Forward, flyback, push-pull, half-bridge, full-bridge, single active bridge, and dual active bridge converters are the different isolated dc–dc converter topologies. Figure 4.13 shows a dual active bridge dc–dc converter. As the name suggests, the input and output supply for any dc–dc converter is dc. The inclusion of a transformer can allow for the stepping up or stepping down of the voltage in addition to the isolation between the input and output. However, a transformer works on ac supply; hence, the input dc supply is converted to ac and can be stepped up or down as per the requirement. With rectification, the secondary voltage can then be converted to dc. The barrier in terms of buck factor or boost factor is removed with the transformer. The frequency of the ac supply involved is in the range of kilohertz. Hence, high-frequency ferrite cores are employed in transformers.

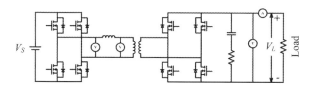

Figure 4.13 Dual active bridge dc–dc converter.

4.4.3 dc–ac converters

Electric drives are widely used in industries. Electric drives involve efficient control of the flow of power with the power converter from the supply to the ac motor so that the speed torque characteristics of the electric motor match what is demanded by the load. Predominantly ac motors are employed in electric drives due to the numerous advantages of ac motors over dc motors. Besides industrial operations, electric drives are employed in electric vehicles, electric aircraft, air-conditioners, refrigerators, washing machines, hard disk drives, pumping units, etc. Electric drives provide variable voltage variable frequency supply at the stator terminals. The grid frequency is fixed at 50 Hz or 60Hz. In order to facilitate the availability of a variable voltage variable frequency supply, the ac supply is converted to dc utilizing rectifiers. This dc supply is then converted to a variable voltage variable frequency ac supply with dc–ac converters, commonly known as inverters. Inverters are universally employed for dc–ac conversion in electric drives, as well as uninterruptible power supplies, dual active bridge dc–dc converters, the grid integration of renewable energy, etc. The power circuit of a single-phase full-bridge inverter, shown in Figure 4.14, consists of four semiconductor switches that have antiparallel diodes. MOSFETs are employed in inverters for low power applications. For high and medium power applications, IGBTs are employed in inverters. In full-bridge inverters, the dc source is connected at input terminals. The two legs of the inverter are formed by a series connection of S_1–S_3 and S_2–S_4. These inverter legs are connected between the positive and negative dc rails. Dedicated control is required for the switching of semiconductor switches in order to obtain ac output voltage. To obtain a positive half cycle, switches S_1 and S_4 are turned ON, keeping other switches in the leg open to avoid a short circuit across the leg. Similarly, in order to obtain negative voltage, switches S_3 and S_4 are turned ON, keeping switches S_1 and S_4 open. Such control of inverter switches with appropriate timings results in square wave ac output voltage output at the ac terminals. To obtain sinusoidal currents, sinusoidal pulse-width modulation can be employed. By adding one more leg to the inverter circuit shown in Figure 4.14, it can be extended to three-phase loads.

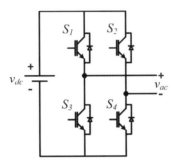

Figure 4.14 Full-bridge inverter.

Based on whether the source is a voltage source or current source, an inverter can be classified as a voltage source or current source inverter. Also, depending on the number of output phases, the inverter can be classified as a single-phase or three-phase inverter. Furthermore, depending on the level of output voltage, they can be classified as 2-level, 3-level, 5-level, 7-level, 9-level, 11-level inverters. Multilevel inverters can be classified not only by level but by the circuit configuration as (i) neutral point clamped or diode clamped, (ii) flying capacitor, or (iii) cascaded configuration.

4.4.4 ac–ac converters

ac–ac conversion can be further classified as:

(i)   ac voltage regulators or ac voltage controllers.
(ii) Cycloconverter.
(iii) Matrix converter.

Figures 4.15–4.18 show the power circuit configuration for an ac voltage controller, cycloconverter, direct matrix converter and indirect matrix converter, respectively. An ac voltage controller controls the RMS value of the output voltage. The frequency of the output voltage is the same as that of the input. Antiparallel SCRs or TRIACs are employed in ac voltage controllers. Cycloconverters employ SCRs and can control the RMS as well as the frequency of the output voltage. As there is no control over the instant of turn OFF of an SCR, forced commutation is required when frequency step-up operation is implemented. Moreover, the currents at the input and output of a cycloconverter are highly distorted. A matrix converter can perform an ac–ac conversion without the need for energy storage elements and provides control over the RMS value of the voltage as well as the frequency of the output supply. Direct and indirect matrix converter topologies have been reported. A matrix converter requires bidirectional switches, which are usually developed with the common emitter connection of IGBTs. Protection of these switches is a key concern in matrix converters.

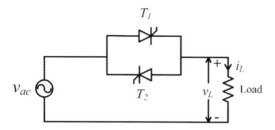

Figure 4.15 Single-phase ac voltage controller.

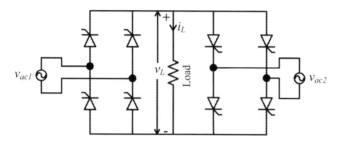

Figure 4.16 Cycloconverter.

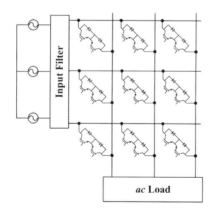

Figure 4.17 Direct matrix converter.

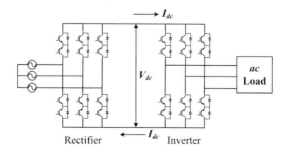

Figure 4.18 Indirect matrix converter.

4.5 ROLE OF POWER CONVERTERS IN RENEWABLE ENERGY SYSTEMS

Power semiconductor switches are the basic building blocks of power converters. Different types of power semiconductor switches in different ratings are available in the market. Power semiconductor switches are selected based on the power converter and the requirements of its operation. Power converters effectively modulate the flow of power. Renewable energy sources, such as wind and solar energy, are more often than not dependent on atmospheric conditions. Hence, the generated power varies with the atmospheric condition. However, when supplying electric power to the grid or to a load, a certain level of voltage and frequency need to be maintained as per the specifications. Power converters play a significant role in controlling the flow of power from a PV panel or a wind turbine to the grid or load. The power converter operation ensures that the power is delivered as per the specifications. This topic is discussed in detail in the next chapter.

REFERENCES

Batarseh, I., & Harb, A. (2018). *Power Electronics Circuit Analysis and Design*, 2nd ed, Springer.
Bimbhra, P. S.(2014). *Power Electronics*, 5th ed. Khanna Publishers, Delhi.
Bose, B. K. (2001). *Modern Power Electronics and AC Drives*, Prentice Hall.
Hart, D. W. (2017). *Power Electronics*, McGraw Hill Education.
Mohan, N., Undeland, T. M., & Robbins, W. P. (1995).*Power Electronics Converters, Applications, And Design*, 2nd ed. John Wiley, New York.
Rashid, M. H. (2012). *Power Electronics Circuits, Devices, and Applications*, 3rd ed, Pearson.
Wu, B. (2006), *High-Power Converters and ACDrives*, Wiley–Blackwell.

Chapter 5

Grid integration of renewable energy sources for buildings

Amit Vilas Sant, Arpit J. Patel, and V.S.K.V. Harish

CONTENTS

5.1 RENEWABLE ENERGY SOURCES

The economic growth and quality of life in any county are chiefly dependent on the supply of electrical energy. Moreover, the consumption of electrical energy and fossil fuels is interconnected with the gross domestic product. Coal-fired thermal power plants are largely responsible for electric power generation. The tremendous growth in global population has resulted in an ever-increasing demand for electric energy. As a consequence, more and more thermal power plants have been established, which results in increased greenhouse gas emissions. The tremendous growth in the petrol and diesel-based transportation sector has further added to the problem of greenhouse gas emissions.

DOI: 10.1201/9781003211587-5

Thermal power plants and the transportation sector, respectively, account for a major portion of greenhouse gas emissions. These greenhouse gas emissions are mainly responsible for global warming and climate change. Due to the melting of polar ice, there is an existential crisis for many islands and coastal regions. Furthermore, the depletion of fossil fuel threatens to derail the growth and development of any society. This has forced governments and various organizations to explore alternate measures such as renewable energy and electric vehicles.

Renewable energy offers a cleaner alternative for electric power generation with the potential to meet the ever-increasing electric power demand. There is an increasing focus on the use of renewable sources, such as solar, wind, tidal, and geothermal energy, for electric power generation. Different countries have put various policies in place to encourage the usage of renewable energy. For example, a subsidy was provided by the Government of Gujarat, India, for the installation of rooftop solar in the domestic sector. The increasing awareness regarding global warming and climate change has further promoted the use of renewable energy. Advancements in material science, power and signal electronics, communication technology, and computing solutions have added to the feasibility for the deployment of renewable energy-based electric power generation.

The clean and abundant availability of wind and solar energy, combined with the maturity of technology, has resulted in their growing usage for electric power generation. Table 5.1 shows the contribution of hydro, solar, wind, geothermal, tidal energy, and other sources for the generation of electric power as per the data presented in the Renewable Capacity Statistics 2021 report by the International Renewable Energy Agency. There is a significant contribution from hydropower, but with geographical and political constraints, its deployment rate is almost saturated. As per this report, the global installed capacity of wind and solar energy has touched 733.276 GW and 713.97GW, respectively. The Global energy transformation: A roadmap to 2050 (2019 edition) report states that with the cost of solar and wind-based generation reducing gradually, this capacity is envisaged to touch 6000 GW and 8500GW by 2050. These energy sources are very much affected by atmospheric conditions. The resulting intermittency in the supply of electrical energy is one of the key concerns. This can be overcome with battery backup, but it increases cost and needs timely replacements. On the other hand, the grid integration of renewable energy systems offers an attractive solution that does not necessitate battery backup, ensures reliability of supply, and provides the provision for reduced annual energy costs with bidirectional power flow. This has given a great impetus to the wider acceptance of solar and wind energy-based electrical power generation. As per the data presented in the Renewable Capacity Statistics 2021 report by the International Renewable Energy Agency, Figure 5.1 shows the yearly total installed capacity of grid-tied renewable energy sources for the past decade. The global share of renewable energy-based generation has increased at a tremendous rate in the past decade. Moreover, based on the same report, Figure 5.2 shows the percentage renewable energy share of electricity capacity. Government policies involving different incentives for the residential, commercial, and industrial sectors for the installation of solar and wind energy systems at their premises have given impetus to their wider acceptance.

Total Renewable Energy (Grid Tied Energy Systems) in MW

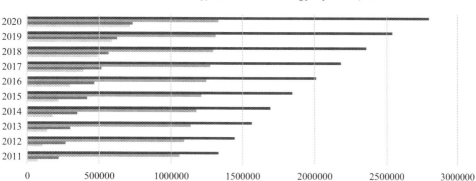

Figure 5.1 Year-wise total installed capacity of grid-tied renewable energy sources for the past decade.

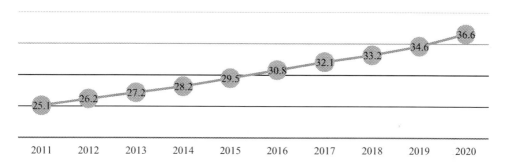

Figure 5.2 Percentage renewable energy share of electricity capacity.

The generation of electrical energy is largely focused in terms of greenhouse gas emissions. However, the transmission, distribution, and consumption of electrical energy also need to be equally focused. Energy savings for increasing transmission, distribution, and load efficiency can have a significant impact on reducing greenhouse gas emissions. As per the statistics presented by Hayter and Kandt, nearly 40% of the global annual energy is consumed by buildings for lighting, heating, air conditioning, cooking, entertainment, etc. Considering the rise in global energy consumption of 1.4% per year, by 2035, buildings will be responsible for the consumption of 296 quadrillion Btu. As per the 2020 Global Status Report for Buildings and Constructions published by the United Nations, residential and non-residential buildings directly and indirectly account for 28% of emissions. A reduction in the energy consumption of a building, which can result in reduced emissions, can be achieved through the usage of energy-efficient equipment and components and incorporating renewable energy-based electric power generation to meet energy requirements. The use of renewable energy to power buildings would involve the localized generation, distribution, and consumption of electrical energy, as a consequence of which transmission and distribution losses in the traditional network can be appreciably reduced. Moreover, energy security concerns, energy conservation, reliability of power supply, and annual energy bills can be positively impacted. Furthermore, with multifunctional control, renewable energy systems can increase the power transfer capability and efficiency of the network and reduce grid pollution by ensuring reactive power compensation and harmonic current mitigation.

Table 5.1 Percentage share of the individual renewable energy sources to that of total renewable energy

Energy source	% Share of the total renewable energy
Renewable hydropower	41.20
Solar photovoltaic	25.28
Onshore wind energy	24.97
Solid biofuel	3.13
Mixed hydro plants	2.05
Offshore wind energy	1.23
Biogas	0.72
Renewable municipal waste	0.55
Geothermal energy	0.50
Solar thermal energy	0.23
Liquid biofuels	0.13
Marine energy	0.02

5.2 RENEWABLE ENERGY SYSTEMS

The electrical power generated from a renewable energy source usually varies with different operating conditions. In a solar energy system, photovoltaic (PV) panels are responsible for the conversion of solar energy into electrical energy. In a wind energy conversion system, wind turbines and an electric generator are responsible for the conversion of wind kinetic energy into electrical energy. The electrical power output of a PV panel depends on the irradiance level and ambient temperature, whereas the electrical output of the wind turbine is dependent on the velocity of the wind. Thus, variable power output is obtained with solar and wind energy systems. Moreover, the current and voltage levels of the generated power may not be as per the standard specifications. Hence, power converters are required to modulate the flow of generated power to the load or grid. Renewable energy systems deployed for electric power generation can be categorized as standalone or grid-tied. Standalone systems involve harvesting renewable energy for the generation of electric power to be fed to an ac or dc load without any involvement of the grid. Such a system necessitates battery backup for ensuring uninterrupted supply to the load at all times. On the other hand, the grid-tied system involves the interconnection of a renewable energy system, load, and grid. This point of interconnection is termed the point of common coupling (PCC). The modulated ac power is fed to the load, and if excess power is available it is supplied to the grid. If the generated power is not able to meet the demand, then power can be drawn from the grid. Due to its obvious advantages, such as the absence of battery backup, reliability of supply, and scope for power trading, the grid-tied system finds greater application. The grid integration of a renewable energy source necessitates adherence to certain codes, generally known as grid codes, so as to meet the operational and functional requirements at the point of interconnection in normal as well as adverse grid conditions.

5.3 IEEE STANDARD 1547

In the energy domain, concern is increasing towards the deployment of renewable energy resources to meet increased demand with minimal environmental impact. This leads to a requirement for the interconnection of renewable energy resources with electrical power systems. IEEE Standard 1547 details the operational and functional criteria for the interconnection of distributed energy sources such as solar PV systems. Interconnecting electric power systems are defined in IEEE Standard 1547. Grid integration of solar PV and wind energy systems must meet the technical requirements specified in IEEE Standard 1547. IEEE Standard 1547 is useful for a wide range of organizations dealing with electric energy, including electric utilities, state electric regulating or policymaking entities, and solar and other DER developers, integrators, and installers. This standard describes the technical requirements related to the following major aspects.

- Interconnection and interoperability
- Operation
- Power quality
- Maintenance
- Testing
- Safety
- Security

Different definitions related to grid integration are mentioned below.

Distributed energy resource (DER): Electric power generation units (such as photovoltaic panels and wind generators) and energy storage elements that need power electronic interface for connected to the power system.

Interconnection equipment: All devices employed in an interconnection system.

Interconnection system: All interconnection equipment and operations employed for interconnecting a single or or several DER units to an area's electric power system.

Point of DER connection: The point at which DER is electrically connected to the electric power system while fulfilling the conditions set by the standard.

Total demand distortion (TDD): The percentage ratio of the roost of the sum of the square of harmonic components to the load current corresponding to the maximum demand.

Total harmonic distortion (THD): The closeness of a wave shape to its fundamental component can be mathematically expressed as the percentage ratio of the root of the sum of harmonic currents to its fundamental.

Total rated current distortion (TRD): The ratio of the root of the sum of the square of the harmonic currents measured at the DER terminals, feeding a linear load, to the greater of the total current demand or DER capacity, expressed in percentage.

The primary requirement is that the DER and electric power system grounding should be the same. The synchronization limits for the voltage, frequency, and phase angle are mentioned when a DER is operated in interconnected mode.

For the grid integration of a DER, the operational requirements are stated below:

- DER output voltage should match grid voltage.
- DER output frequency should match grid frequency.
- Phase angle of DER output voltages should match that of the respective grid voltages.

The maximum limit for step increase in active power is also mentioned. While integrating solar PV systems with the grid, a 5% current THD limit has to be maintained. Moreover, there is a limit to the distortion caused by the individual harmonics. As per IEEE 1547–2003, the DER should not actively regulate voltage and should be tripped on the abnormal voltage/frequency conditions of the grid. However, the revised standard IEEE 1547–2018 states, "*The DER shall provide voltage regulation capability by changes of reactive power. The approval of the Area EPS Operator shall be required for the DER to actively participate in voltage regulation.*" Also, the DER shall be capable of ride-through abnormal frequency/voltage and provide frequency response. This implies that the DER can supply as well as consume reactive power within the given margins. Additionally, the DER should incorporate fault ride-through capability. Furthermore, revised standard IEEE 1547–2018 reinforces power quality aspects in terms of limitations on current distortions, dc current injection, DER-caused voltage fluctuations, overvoltage contribution, and harmonics.

5.4 GRID INTEGRATION OF SOLAR PHOTOVOLTAIC SYSTEMS

5.4.1 Photovoltaic panels

Along with the free and abundant availability of solar energy, advancements in power and signal electronics technology has further led to the ever-increasing usage of solar PV systems. Moreover, the extensive research and advancement in material science for increasing the efficiency of solar cells has definitely contributed to the popularity of solar PV technology. With recent developments in technology, rooftop solar installations are a common feature in residential, commercial, and industrial areas. Government incentives and public awareness have encouraged solar rooftop installations. Solar PV technology, a significantly matured technology, involves the conversion of solar energy into electrical energy. The widely used material for the development of a basic PV cell is a crystalline silicon that is relatively thicker than the new thin-film cells made up of amorphous silicon and non-silicon material such as cadmium telluride. The thickness of a PV cell made up of crystalline silicon and a thin-film cell has corresponding thickness in the order of 200–500 μm and 1–10 μm. The amount of material required by thin-film cells is comparatively less than that required by crystalline silicon cells. However, first-generation thin-film cells have an efficiency of about 50% of that obtained with conventional thick silicon cells. Thin-film cells are also comparatively less reliable over time.

Solar radiations with a wavelength shorter than 1.11 μm are required for the generation of free electrons. If shorter length radiations are imposed on silicon cells, the extra energy will be wasted in the form of heat inside the cell. When solar radiations with the appropriate wavelength are absorbed by the PV cell, electron and hole pairs are formed. The electrostatic force of the depletion layer pushes the electrons into the n region and hole into the p region. This accumulation of hole and electrons correspondingly in the p and n regions gives rise to the voltage. This voltage can drive the current through the load connected across the terminals of the PV cell. The voltage developed across the PV cell is around 0.5 V, which is not sufficient for most applications. A series connection of PV cells results in an increase in voltage level, and the parallel connection of PV cells can result in an increase in current being delivered. The resulting PV module, which operates at higher voltage and current levels, forms the basic building block for PV applications. Further, PV arrays are formed by the series and parallel connection of PV modules. The series connection of y number of PV modules results in voltage across the series connection being y times the voltage of the individual module. Such series connection of modules is termed as a string. The parallel connection of such z number of strings results in the current delivered being z times the current being delivered by the individual module. In order to increase the current and voltage rating, modules are connected in series and parallel configuration to construct a PV array having a rated voltage and current as per the requirement of the application. The pictorial illustration of a PV panel, and the I-V and P-V characteristic curve of a PV module, are shown in Figure 5.3. The terminal voltage of a PV module when the terminals are open-circuited is called open-circuit voltage, V_{oc}, and the current supplied by the PV module when its terminals are short-circuited is referred to as short circuit current, I_{sc}. A power versus voltage curve is included in the same figure. The output power is the product of the voltage and the current. The power during the open circuit and short circuit is zero as the voltage and current are zero in respective cases. The point near the knee of the I-V curve at which the product of the voltage and the current (i.e., the power) is the maximum is termed the maximum power point (MPP). The voltage and current corresponding to the MPP under standard test conditions (STC) are mentioned in datasheets. The STC is characterized as the solar insolation of 1000W/m², air mass ratio of 1.5, and cell temperature of 25°C. To evaluate the performance of a PV module, a fill factor is used, which is defined as a ratio of the power at MPP to the product of V_{oc} and I_{sc}. The PV curve shifts according to the atmospheric conditions mainly due to variations in the solar insolation and temperature. The current and output power of the PV panel are highly dependent on the solar insolation. The output current at given condition is given as:

$$I_{out} = Solar\ Insolation \times \frac{1000}{\left(\begin{array}{c} Short\ circuit\ current \\ corresponding\ to\ the \\ isolation\ level\ of\ 1000\ W\ /\ m^2 \end{array} \right)} \tag{5.1}$$

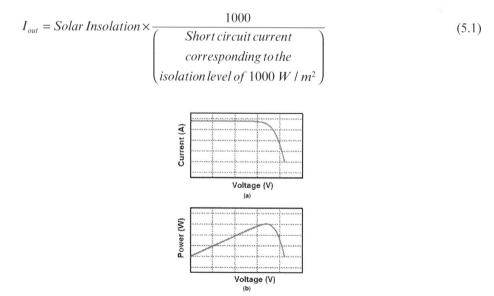

Figure 5.3 I-V and P-V curves of a PV panel.

The increase in cell temperature also causes a reduction in the output power supplied by the PV array. The operating point of the PV array depends on atmospheric and loading conditions. To ensure the supply of maximum permissible power from the PV module, maximum power point tracking (MPPT) algorithms are implemented. This algorithm maintains the operating point of a PV array at the MPP, which results in maximum power being extracted from the PV array. MPPT algorithms are discussed in the following section.

5.4.2 Mathematical modelling of photovoltaic panel

Based on the power structure,grid-tied PV systems can be classified as dual or single-stage systems. A PV cell can be represented by a simple equivalent circuit, as shown in Figure 5.4. A simple equivalent circuit of a PV cell is represented by a current source having a magnitude I_{SC} and a parallel diode, D. The output current equation is stated as:

$$I_{out} = I_{SC} - I_d \qquad (5.2)$$

where I_{out} is the output current of a PV cell, I_{SC} is the ideal source current (i.e., equal to the short circuit current of a PV cell), and I_d is the diode current.

The Shockley diode equation can be used to define the relationship between the current and voltage for a diode as:

$$I_d = I_o \left(e^{(qV/kT)} - 1 \right) \qquad (5.3)$$

where I_o is the reverse saturation current of the diode, V is the voltage across the diode, q is the charge of electron equal to 1.602×10^{-19}C, k represents Boltzmann's constant (i.e., equal to 1.381×10^{-23}J/K), and T is the junction temperature in K.

On substituting equation (5.3) in equation (5.2), the output current equation is obtained as:

$$I_{out} = I_{SC} - I_o \left(e^{(qV/kT)} - 1 \right) \qquad (5.4)$$

The open-circuit voltage equation is measured by keeping the terminals of the PV cell open. Open circuit voltage V_{OC} is calculated by keeping the output current I_o equal to zero, as stated in equation 5.5.

$$V_{OC} = \frac{kT}{q} \ln \left(\frac{I_{SC}}{I_o} + 1 \right) \qquad (5.5)$$

Using the above modelling equations, the I-V curve of the PV cell can be plotted.

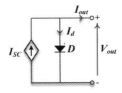

Figure 5.4 Equivalent circuit of a PV cell.

When cells are connected in series and one of the series-connected cells is not generating any current, the simple equivalent circuit infers that no current is obtained at the output. In a practical scenario, some current is obtained at output even if one of the series connected cells is not generating any current. This physical phenomenon can be included in an equivalent circuit by connecting parallel resistance R_{sh} in a simple equivalent circuit. Further, series resistance R_{sh} is included to model resistance due to contact between a cell and its leads and the resistance offered by the semiconductor layers. The practical equivalent circuit of a PV cell is represented in Figure 5.5.

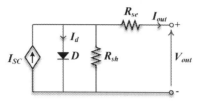

Figure 5.5 Practical equivalent circuit of a PV cell.

Considering series and parallel resistance in the equivalent circuit, the output current is stated as:

$$I_{out} = I_{SC} - I_o \left[\exp\left[\frac{q\left(V_{out} + I_{out}R_{se}\right)}{kT} \right] - 1 \right] - \left(\frac{V + IR_{se}}{R_{sh}} \right) \tag{5.6}$$

The above equation represents the relationship between output current and voltage, which can be used to model a practical PV cell.

5.4.3 Maximum power point tracking algorithm

Practically, the reported maximum efficiency of a PV panel is found to be around 15%. Moreover, solar radiation is only available for around 7–10 hours a day, depending on the weather and location. Hence, it is important that the maximum possible power is extracted from the PV panel. As discussed in section 5.4.1, if the voltage across the terminals of the PV array or current being supplied by the PV array is maintained at the value corresponding to the MPP, then the maximum power can be harvested. This MPP shifts as per the irradiance level and temperature of the PV panel. To track the corresponding values of current or voltage with respect to the MPP, MPPT algorithms are employed. The perturb and observe (P&O) method, also known as the hill-climbing method, is a very popular MPPT algorithm. The incremental conductance (IC) algorithm is another popularly employed method. Besides this, the open-circuit voltage method and fuzzy and ANN-based techniques are also reported for MPPT. The flow charts for the P&O and IC algorithm are shown in Figures 5.6 and 5.7, respectively. In these flowcharts, the terminal voltage of the PV panel is denoted as $V(t)$, the current supplied by the PV panel is denoted as $I(t)$, the power supplied by the PV panel is denoted as $P(t)$, time is denoted as t, time step is denoted as Δt, the difference between the present and previous values of voltage and power associated with the PV array are respectively denoted as $V(t-1)$ and $P(t-1)$, the voltage to be maintained across the terminals of PV array for harvesting maximum power is denoted as V_{ref} and the value by which V_{ref} is varied is denoted as ΔV.

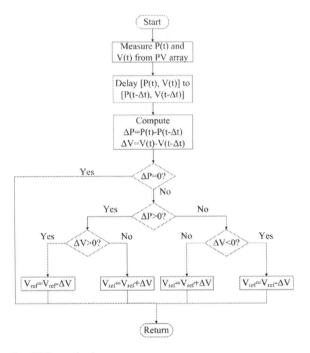

Figure 5.6 Flowchart for P&O method.

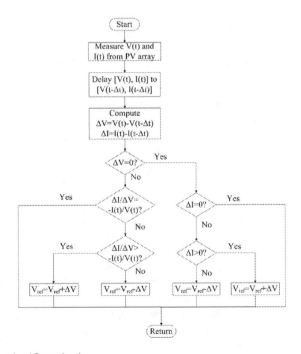

Figure 5.7 Flowchart for IC method.

5.4.4 Power structure

Depending on the power structure,grid-tied PV systems can be classified as dual or single-stage systems. In a dual-stage system, the cascade connection of a dc–dc converter and a voltage source inverter (VSI) are employed to interface the PV array with the grid. However, only a VSI is employed to interface the PV array with the grid in a single-stage system. Figures 5.8 and 5.9 show the block diagram representation of the single-stage and dual-stage grid-tied PV system, respectively. In a dual-stage system, a dc–dc converter is employed to ensure that voltage corresponding to the MPP for the given operating condition is maintained across the PV terminals. The VSI operates to regulate the dc link voltage, control the flow of power from the PV panel to the grid, and ensure that the currents are injected at a unity power factor with respect to the respective phase voltage at the PCC. The operation of a VSI also ensures grid synchronization and adherence to the grid codes. Grid synchronization is the process of connecting the incoming distributed resource to the grid while ensuring that the phase, frequency, and voltage of the incoming distributed resource match with the grid. In a single-state system, besides all the mentioned functionalities of the VSI, the VSI also ensures that voltage across the PV array is regulated at the value corresponding to the MPP for the given operating condition. Due to the extensive usage of nonlinear loads, the current and voltages are often distorted. Hence, the power factor is considered as the product of the displacement factor and the distortion factor. Regulating the power factor of the injected currents at unity implies that only active power is injected by the PV system and no reactive power is injected. This is in accordance with the established grid codes. Due to the reduced number of power electronic switches and the consequent reduction in the gate driver and associated signal electronics, the single-state system has a lower component count and is more reliable. Moreover, with a lower number of power electronic switches, switching and conduction losses are fewer; hence operation is more efficient. Additionally, a lower switch count and losses also result in a smaller size of heat sink and a more compact system.

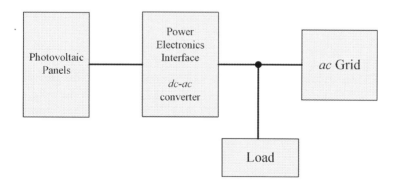

Figure 5.8 Block diagram representation of the power structure of a grid-tied single-stage solar PV system.

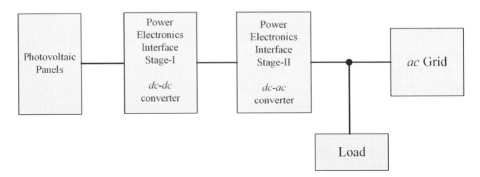

Figure 5.9 Block diagram representation of the power structure of a grid-tied single-stage solar PV system.

The detailed power structure of a single-stage grid-tied PV system is shown in Figure 5.10, where v_{pv} is the voltage across the terminals of the PV array, i_{pv} is the current being supplied by the PV array, D_B is the blocking diode, C_{dc} is the capacitor connected at the dc link, v_{dc} is the instantaneous value of the dc link voltage, Q_1-Q_6 are the IGBT switches that constitute the three-phase VSI, G_1-G_6 are the respective gating signals for Q_1-Q_6, L_x is the coupling inductor that blocks the propagation of high-frequency current harmonics into the grid, C_r is the capacitor that acts as a ripple filter to prevent the high-frequency component of voltage from distorting the grid voltage, v_{Ga}-v_{Gb}-v_{Gc} are the instantaneous value of phase voltages corresponding to phases *a-b-c*, i_{Ga}-i_{Gb}-i_{Gc} are the value of instantaneous currents being supplied by the grid, $(R_G + jX_G)$ is the per phase line impedance, v_{Pa}-v_{Pb}-v_{Pc} are the value of instantaneous phase voltages corresponding to phases *a-b-c* as measured at the PCC, i_{La}-i_{Lb}-i_{Lc} are the value of instantaneous currents being consumed by the load, and i_{Xa}-i_{Xb}-i_{Xc} are the value of instantaneous currents being supplied by the grid-tied PV system.

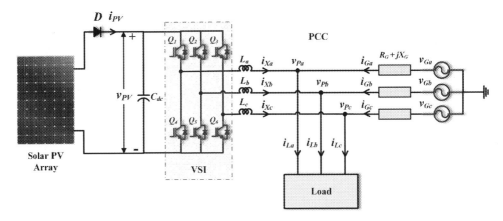

Figure 5.10 Single-stage grid-tied PV system.

Voltage sensors are deployed for the measurement of instantaneous values of v_{dc} and v_{Pa}-v_{Pb}-v_{Pc}. It should be noted that v_{dc} equals v_{pv}. Similarly, current sensors are employed for the instantaneous values of i_{Ga}-i_{Gb}-i_{Gc}, i_{La}-i_{Lb}-i_{Lc}, and i_{Xa}-i_{Xb}-i_{Xc}. For the MPPT algorithm, i_{pv} is also measured. These measured values serve as an input to the controller, which utilizes them for the implementation of the control algorithm. The measured values undergo analogue-to-digital conversion (ADC) in the controller at the rising edge of the start of the conversion signal. After the completion of each sampling duration, the start of the conversion signal goes high for a specific duration to initiate the ADC operation. The reference voltages generated by the algorithm serve as an input to the hardwired pulse-width modulation (PWM) logic available on the embedded controller board. Counters are used to generate the high-frequency carrier signal for the PWM logic. Based on the reference and carrier signal, the hardwired PWM logic generates the gating signals G_1-G_6. The reference signal may be sinusoidal or sinusoidal with the presence of third harmonics depending on whether a sinusoidal PWM or a space vector PWM is employed. Agate driver is used to interface the PWM pins of a low power controller with the gating terminals of a VSI. The gate driver provides necessary optical isolation and voltage bias. Moreover, functionalities such as shoot-through protection and boost strapping are also provided. This control over the ON/OFF duration of Q_1-Q_6, which constitute the three-phase VSI, ensures grid synchronization of the solar PV system. Also, the flow of power from the PV array to the grid is controlled so that only active power is supplied to the ac grid while the injected reactive power is held at zero. This is as per the rules set by the grid codes. In addition to this, dc-link voltage is controlled so that the terminal voltage of the PV panel is maintained at the value corresponding to the MPP for the given ambient temperature and irradiance levels. With the extensive usage of nonlinear loads, the propagation of current harmonics, consequent voltage harmonics, and resulting power quality degradation is also a matter of huge concern. The grid integration of a solar PV system – or any distributed resource – should not cause a degradation in power quality. In order to ensure this, the inverter is controlled to inject currents with THD levels within the specified levels. Such control of the VSI results in IEEE1547 compliant operation of the grid-tied PV system with the maximum power being harvested from the PV array.

5.4.5 Voltage-oriented control

Voltage-oriented control (VOC) is a control algorithm widely employed for the control of grid-tied inverters. VOC draws its inspiration from the field-oriented control (FOC) of ac motors. While FOC implements the decoupled control of d-q axes currents representing flux and torque producing components of current, VOC implements decoupled control of d-q axes currents corresponding to active and reactive power being injected into the grid. The structure of FOC involves an outer speed control loop and two inner d-q axes current control loops, whereas the structure of VOC comprises an outer dc-link voltage control loop followed by two inner d-q axes current control loops. FOC involves abc-to-$dq0$ and $dq0$-to-abc transformation. Likewise, VOC also involves these transformations. In FOC, instantaneous rotor position information is required for the implementation of the two transformations. Similarly, in VOC, the information regarding the instantaneous phase angle of the fundamental positive sequence component of PCC voltages is essential for the transformations as well as control.

Figure 5.11 shows the block diagram representation of VOC for the grid integration of a solar PV system. MPPT algorithms process i_{pv} and v_{pv} to estimate v_{mpp}, the voltage to be maintained across the terminals of a PV array for harvesting the maximum possible power under the given operating condition. The outer dc-link voltage control loop involves the comparison of v_{dc} with v_{mpp} and the processing of the resulting error through a dc-link voltage controller, which can be a proportional-integral (PI) controller. PI controllers are widely preferred due to their computational simplicity. Tuning the controller gain determines its steady-state and dynamic response. The dc-link voltage controller determines the d-axis current to be supplied, which would result in v_{dc} being regulated at v_{mpp}. This implies that v_{pv} is maintained at v_{mpp} and thereby results in the operation of the PV array at the MPP point and the harvesting of maximum power. The inner d-q axes current control loops individually process the difference between the reference and actual values of the d-q axes currents through current controllers. The reference d-axis current, i_{dREF}, is the output of the voltage control loop and is proportional to the active power to be supplied by the PV system. In the absence of the need for reactive power injection, the q-axis reference current, i_{qREF}, is kept at zero. The actual d-q axis currents, i_d-i_q, are determined when i_{Xa}-i_{Xb}-i_{Xc} undergo abc-to-$dq0$ transformation. It should be noted that i_d-i_q are respectively proportional to the active and reactive power being supplied by the PV array to the ac grid. Also, the d-q axes currents are dc quantities. Thus, instead of controlling the phase, frequency, and magnitude of ac currents, with the transformations, VOC controls the magnitude of equivalent dc quantities with phase and frequency being maintained as per the respective grid voltage. i_{dREF}-i_{qREF} are correspondingly compared with i_d-i_q to determine the d-q axes current errors. The two d-q axes current controllers process the respective current errors for computing the reference d-q axes voltages, v_{dREF}-v_{qREF}. The output of VSI is ac quantity, whereas v_{dREF}-v_{qREF} are the dc quantities. Hence, v_{dREF}-v_{qREF} are transformed into a-b-c reference frame with $dq0$-to-abc transformation. The reference voltages in a-b-c reference frame v_{aREF}-v_{bREF}-v_{cREF} are compared with the high-frequency carrier signal to generate the gate signals, G_1-G_6. The resulting switching action ensures the injection of requisite active power into the grid with no reactive power being injected. VOC-based control of a VSI results in i_{Xa}-i_{Xb}-i_{Xc} having a frequency that matches with the grid frequency, ω, and phase angle matching that of v_{Pa}-v_{Pb}-v_{Pc}, respectively. The use of $\omega_1 t$, where t is time, for abc-to-$dq0$ and $dq0$-to-abc transformation results in phase and frequency of i_{Xa}-i_{Xb}-i_{Xc} matching the respective fundamental positive sequence component of v_{Pa}-v_{Pb}-v_{Pc}. A phase-locked loop (PLL) processes v_{Pa}-v_{Pb}-v_{Pc} to determine $\omega_1 t$. PLL-based synchronous reference frame theory and second-order generalized integrators are employed for determining $\omega_1 t$.

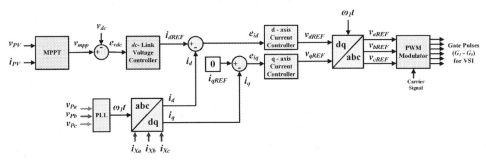

Figure 5.11 Block diagram representation of VOC for the grid integration of a solar PV system.

5.5 GRID INTEGRATION OF WIND ENERGY CONVERSION SYSTEMS

5.5.1 Wind energy

The wind energy system provides an attractive solution for adopting renewable energy resources with zero carbon emissions. Wind energy has been harvested forages, either through the sails of ships or windmills. Since the advent of large-scale electric power generation for cities, towns, and villages, prime mover-driven electric generators are responsible for generation irrespective of the type of power plant. Wind energy could also drive the prime mover to facilitate the necessary motion for electric power generation. However, with wind velocity being variable, it would result in generated voltages at the output of the electric generator having variable frequency and amplitude. This would be unsuitable as per the existing standards of grid integration. With the commercialization and development of power electronic technology and signal processors, the flow of power from these generators can be regulated with power electronic converters while adhering to the grid codes. With this knowledge, the usage of wind energy conversion system-based electric power generation has grown by leaps and bounds. According to the latest IRENA figures, installed wind-generation capacity onshore and offshore has expanded by approximately 75 times in the past two decades, from 7.5 GW in 1997 to 564 GW in 2018. Between 2009 and 2013, wind energy output doubled. Although there are strong winds in numerous locations across the globe, the prime locations for generating power from wind are mostly in remote areas. Offshore wind turbines offer great potential in this regard. The countries with the highest installed capacity for wind energy-based electric power generation by 2020 are listed in Figure 5.12 based on the data presented in the Renewable Capacity Statistics 2021 report by the International Renewable Energy Agency.

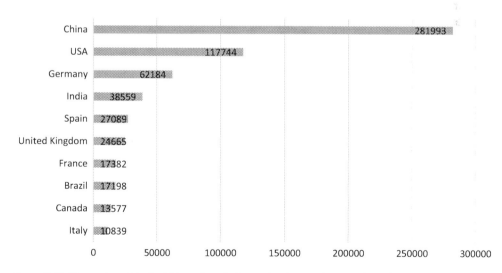

Figure 5.12 Countries with the highest installed capacity for wind energy-based electric power generation by 2020.

A wind energy conversion system involves the conversion of the kinetic energy (KE) of the wind into rotational energy, which is further utilized for rotation of the generator shaft so that the mechanical energy can be converted into electrical energy and be made available at the stator terminals of the electric generator. The KE of the air having mass m, travelling at velocity, v, is given as:

$$KE = \frac{1}{2}mv^2 \tag{5.7}$$

Further, since power is defined as energy per unit time, the power generated by a mass of air, P_a, travelling through an area A at velocity v is represented by:

$$P_a = \frac{\frac{1}{2}mv^2}{time} = \frac{1}{2}\left(\frac{m}{time}\right)v^2 \tag{5.8}$$

The term $\frac{m}{time}$ represents mass flow rate, which is equal to ρAv, where ρ is the density of air.

On substituting the term $\frac{m}{time}$ with ρAv, wind power can be expressed as:

$$P_a = \frac{1}{2}\rho Av^3 \tag{5.9}$$

As per SI units, v is the wind speed normal to A in m/s, A is the area swept by rotor blades in m^2, and P_a is the power in the wind in watts. In this relationship, ρ represents the air density in kg/m^3.

If the wind turbine extracted all the wind energy, then the velocity of the wind after the turbine would drop to zero (i.e., there would be no wind after the turbine). However, in practice, the entire wind energy cannot be harvested. The power coefficient, C_p, needs to be considered, which is the measure of the effectiveness of wind power extraction. C_p is given as:

$$C_p = \frac{Mechanical\ power\ output\ of\ the\ wind\ turbine}{Maximum\ power\ that\ can\ be\ harvested\ from\ the\ wind} \tag{5.10}$$

As a result, turbine power is stated as:

$$P_T = \frac{1}{2}\rho Av^3 C_p \tag{5.11}$$

P_T is always less than P_a. In fact, the maximum extractable power fraction has a theoretical upper limit known as the Betz limit. As per this limit, a traditional wind turbine can harvest power from the wind at a maximum efficiency of 59%, with a highest probable C_p of 16/27. This implies that doubling the wind speed, for example, increases the power eightfold. As a result, when considering wind energy, small increases in wind speed can have a substantial impact on power generation. Increasing the height of the tower and assembling the turbine at the top is one way of ensuring that stronger winds are experienced by the turbine. The friction that the air encounters as it travels over the earth's surface has a significant impact on the wind speed in the first few hundred metres above the ground. The variation in air density is very restricted and hence is not primarily considered. The wind power exhibits a nonlinear quadratic relationship with the radius of the area swept by the wind turbine, emphasizing the benefits of longer wind turbine blades.

5.5.2 Wind turbines

Wind turbines comprise a nacelle mounted on a tower. The nacelle houses low and high-speed shafts, an electric generator, gearbox, brake, yaw mechanism, pitch mechanism, controller, and anemometer. Wind velocity ranges from 10 to 65 miles per hour in wind streams that blow through flat terrains or over-the-hill locations. These wind velocities are good enough for the rotation of wind turbine blades. Rotor blade design parameters such as chord, twist angle, and length are selected to achieve optimum aerodynamic performance and acceleration effects under various wind conditions at different tower height levels. The important components of the wind turbine are mentioned in this section with their functional description. Figure 5.13 shows the detailed structure of a wind turbine. The functionality of each of these converters is stated below:

- **Tower:** The tower is generally made of tubular or lattice steel or concrete and supports the entire structure of the wind turbine. Generally, taller towers are employed to capture more energy from the wind, thereby generating more electricity.
- **Rotor:** Blades and a hub form the rotor of the wind turbine. This facilitates the conversion of the harvested KE of wind into rotational energy.
- **Blades:** The blades rotate when wind is blown, causing the rotor to rotate. Most turbines employ a three-blade structure, considered the most stable mechanical structure with minimal drag force on the blades resulting in minimum vibrations experienced by the structure.
- **Low-speed shaft:** This rotates with the rotation of the blade and is coupled with the gearbox in order to increase the mechanical speed.
- **Gearbox:** This provides mechanical linkage between the low and high-speed shafts while raising rotational speeds to 1,000–1,800 rotations per minute (rpm) from 30–60 rpm. This is to meet the rotational speed required by most generators to produce electrical power. Engineers are working towards "direct-drive" generators that operate at lower rotational speeds and do not require gearboxes because the gearbox is an expensive and heavy component of a wind turbine.
- **High-speed shaft:** This is the shaft coupled to the gearbox and electric generator. It is used to deliver mechanical power to the electric generator
- **Brake:** This is used to stop the rotor in case of emergencies. The brake can be mechanical, electrical, or hydraulic.
- **Generator:** This converts the mechanical power available at the shaft into electrical power. Previously this was usually an induction generator. Nowadays, double-fed induction generators (DFIGs) and permanent magnet synchronous generators (PMSGs) are popular for wind energy systems.
- **Nacelle:** This houses the gearbox, low- and high-speed shafts, generator, controller, brake, etc.
- **Anemometer:** This measures the wind speed and communicates this measurement with the controller for necessary actions.

- **Yaw mechanism:** This consists of a yaw drive and an electric motor. It takes wind direction information from the wind vane so that when the wind direction changes, upwind turbines are oriented to face the wind. In a downwind turbine, the rotor is manually blown away from the wind by the wind itself. Thus, a yaw drive is not required in downwind turbines.
- **Pitch mechanism:** This is used to turn (or pitch) rotor blades away from the wind in order to regulate the rotor speed. It prevents the rotor from rotating in too strong winds or winds that are too weak to generate electrical power.
- **Controller:** This is used to process the various data available from sensors such as wind speed and wind direction and generates command signals for the corresponding drive mechanisms to initiate and regulate operational and functional requirements. Depending on the direction of the wind, the controller controls the yaw motor to spin the rotor towards or away from the wind.
- **Transformer:** A step-up transformer located at the tower's base is used to interface the electrical power available at the output terminals of the power converter with the grid.

The rotational axis around which the turbine blades rotate characterizes wind turbines. For large wind turbines, horizontal axis wind turbines are mainly employed. However, vertical axis wind turbines are more suited for small and medium-sized wind turbines usually employed in buildings. These turbines can be mounted on rooftops as well as the open spaces surrounding the building. While mounting on the rooftop, care needs to be taken that the structure is able to handle the vibration resulting from the operation of the vertical axis wind turbine. Figures 5.14, 5.15, and 5.16 show the horizontal axis upwind turbine, horizontal axis downwind turbine, and Darrieusvertical axis wind turbine.

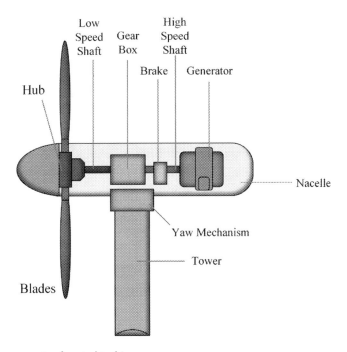

Figure 5.13 Components of a wind turbine.

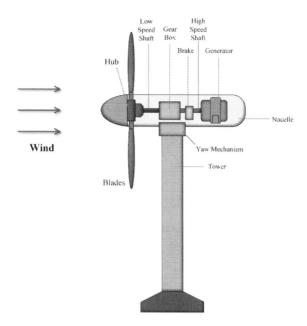

Figure 5.14 Horizontal axis upwind turbine.

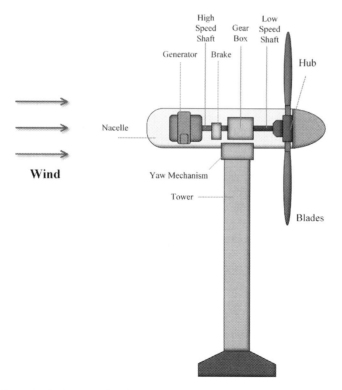

Figure 5.15 Horizontal axis downwind turbine.

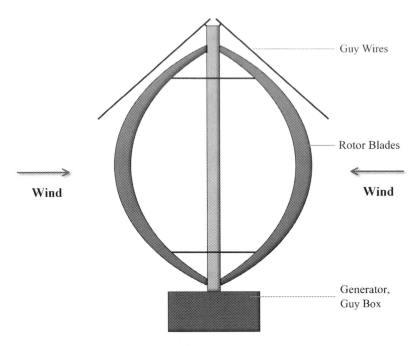

Guy Wires

Rotor Blades

Wind

Wind

Generator,
Guy Box

Figure 5.16 Darrieus vertical axis wind turbine.

5.5.3 Power structure

Wind turbines harvest the KE of the wind and convert it into rotational energy. The low-speed shaft is coupled with the high-speed shaft through the gearbox. The rotation of the low-speed shaft translates to the rotation of the high-speed shat, albeit at a higher speed. This serves as the prime mover action for the electric generator and consequent generation of electrical energy. Due to the lower power density, poorer efficiency, and need for frequent maintenance, dc generators are not usually preferred in this application.

Induction generators and permanent magnet synchronous generators(PMSGs) are used in wind energy applications. PMSGs offer merits of higher efficiency, robust construction, maintenance-free operation, freedom from the need to supply magnetizing current, operation over a wide speed range, etc. These merits make PMSGs ideal candidates for wind energy system applications. The rotor comprises permanent magnets, mainly neodymium iron boron (Nd-Fe-B), mounted on a rotor bore. As the air gap is magnetized by these magnets, no magnetizing current is to be drawn from the stator and the stator currents are supplied at near unity power factor. PMSGs are preferred, including from large to small turbines. Irrespective of the horizontal or vertical axis wind turbine, the control of a PMSG and its integration with the grid is more or less similar.

A type-IV wind energy conversion system, shown in Figure 5.17, comprises a PMSG, where v_w is the velocity of wind, ω_s is the rotor speed, i_{ma}-i_{mb}-i_{mc} are the instantaneous stator currents being supplied by the PMSG, v_{mab}-v_{mbc}-v_{mca} are the instantaneous line-to-line voltages measured across the stator terminals, L_{ma}-L_{mb}-L_{mc} are the inductors interfacing the stator terminals with the ac terminals of the machine end converter, C_{dcw} is the dc-link capacitor, v_{dcw} is the dc-link voltage, L_{wa}-L_{wb}-L_{wc} are the inductors interfacing the ac terminal of the grid end converter with the grid, v_{Ga}-v_{Gb}-v_{Gc} are the instantaneous values of phase voltages of the grid corresponding to phases a-b-c, i_{Ga}-i_{Gb}-i_{Gc} are the values of three-phase instantaneous currents being supplied by the grid, $(R_G + jX_G)$ is the per phase line impedance, v_{Pa}-v_{Pb}-v_{Pc} are the instantaneous value of phase voltages corresponding to phases a-b-c as measured at the PCC, i_{La}-i_{Lb}-i_{Lc} are the values of three-phase instantaneous currents being consumed by the load, and i_{wa}-i_{wb}-i_{wc} are the values of three-phase instantaneous currents being injected by the wind energy conversion system. G_{m1}-G_{m6} are the gate pulses for the machine end converter, whereas Gw7-G_{w1}-G_{w6} are the gate pulses for the grid end converter.

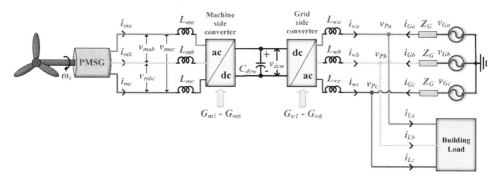

Figure 5.17 Type-IV wind energy conversion system.

In a wind energy conversion system, a wind turbine converts the KE of the wind into mechanical energy. With or without the gearbox, this mechanical energy facilitates the prime mover action that leads to the rotation of the rotor. This action results in the conversion of mechanical energy into electrical energy available at the stator terminals. Variable wind speed results in stator voltages having variable frequency and amplitude. This voltage needs to be regulated before it can be fed to the load or grid. Two-stage power conversion is employed for regulating the voltage and controlling the flow of power from the wind turbine to the grid and load. Two three-phase, six-switch VSIs are connected in a back-to-back configuration with a common dc-link. The power converter connected to the generator is known as a machine end converter, whereas the power converter connected to the grid is called a grid end converter. A necessary control strategy is employed for both the power converters. The machine end converter is controlled with FOC. In FOC, d-q axes current represent the flux and torque producing components of currents, respectively. By controlling these currents, the mechanical parameters of the generator, namely rotor speed and torque, can be regulated. The grid end converter is controlled with VOC to ensure grid synchronization and dc-link voltage regulation. Moreover, the power being fed by the wind energy conversion system into the grid is also controlled. Passive filters are used to interface the grid end converter with the grid. To implement FOC, i_{ma}-i_{mb}-i_{mc}, ω_s and v_w need to be measured with appropriate sensors. To measure speed, an incremental encoder with indexing is very much suited for this application. The rotor position can be measured based on this information. Similarly, for implementing VOC, v_{dc}, v_{Pa}-v_{Pb}-v_{Pc}, and i_{wa}-i_{wb}-i_{wc} need to be measured with the help of current and voltage sensors. The measured values serve as an input to the controller, which utilizes them for the implementation of the FOC and VOC algorithm. The measured quantities are sampled when the sampling duration is elapsed. ADC operation ensures the availability of the digital values if the input signal is available to be processed by the respective algorithms. Quadrature encoder circuitry processes the pulses received from the incremental encoder to determine actual rotor speed and position with the help of appropriate code. FOC is responsible for the control of the machine end converter through the generation of G_{m1}-G_{m6}. On the other hand, VOC is responsible for the generation of G_{w1}-G_{w6} for controlling the grid end converter. The reference voltages generated by each of the two algorithms serve as an input to the hardwired PWM logic available on the embedded controller board. The high-frequency carrier signal, generated by the counter, serves as another input for the PWM logic. Based on the reference signals generated by FOC and VOC and the carrier signal, the hardwired PWM logic generates the gating signals G_{m1}-G_{m6} and G_{w1}-G_{w6}, respectively. For interfacing the PWM pins of the control card with the gate terminals of the IGBTs, gate drives are employed. These gate drives are responsible for ensuring voltage biasing, shoot-through protection, boost strapping, and appropriate voltage biasing. For the grid interfacing of a wind energy conversion system, a grid end converter ensures grid synchronization, dc-link voltage control, control of the flow of power, and IEE-1547 compliant operation. Control of the machine end converter ensures decoupled and energy-efficient control of the PMSG and the harvesting of maximum wind energy. It also ensures that the currents drawn from the PMSG are sinusoidal so that harmonic losses can be eliminated or minimized.

5.5.4 Control of machine end converter

FOC is implemented for the control of a machine end converter. FOC controls the stator current so that the actual rotor speed matches the reference value determined by the MPPT algorithm. As in the case of VOC, FOC also involves *abc*-to-*dq0* and *dq0*-to-*abc* transformations. The block diagram representation of the FOC-based control of a machine end converter is illustrated in Figure 5.18. An incremental encoder with an index pulse serves as the speed and position sensor. With the help of quadrature encoder pulse circuitry, the rotor speed, ω_s, and position, θ_s, are determined. With the help of the measured v_w, the MPPT algorithm estimates the reference rotor speed for PMSG, ω_{sREF}. If the rotor rotates at this optimal reference value, then maximum power can be extracted. ω_{sREF}, when compared with ω_s, provides the speed error, e_ω, which, when considered as an input for the speed controller, results in the estimation of the reference q-axis stator current, i_{mqREF}. i_{mqREF} is proportional to the torque to be generated for ensuring that ω_s matches ω_{sREF}. The permanent magnets mounted on the rotor are responsible for the magnetization of the airgap. The reference value of magnetizing component of stator current is i_{mdREF}. The actual values of the d-q axes components of stator current, i_{md}-i_{mq} are determined by forcing i_{ma}-i_{mb}-i_{mc} to undergo an *abc*-to-*dq0* transformation. Correct estimation of θ_s is required to carry out *abc*-to-*dq0* transformation. i_{md}-i_{mq} are respectively compared with i_{mdREF}-i_{mqREF}. The two PI current controllers respectively process the current errors, e_{md}-e_{mq}, to estimate the reference d-q axes stator voltages, v_{md}-v_{mq}. v_{md}-v_{mq} undergo *dq0*-to-*abc* transformation to obtain the reference stator voltages, v_{maREF}-v_{mbREF}-v_{mcREF}, in an *a-b-c* reference frame. Needless to say, the correct estimation of θ_s is required for carrying out this transformation. If stator phase voltages are maintained at v_{maREF}-v_{mbREF}-v_{mcREF}, then e_{md} and e_{mq} would be zero, implying that i_{md}-i_{mq} are equal to i_{mdREF}-i_{mqREF}, respectively. As i_{mdREF}-i_{mqREF} are proportional to the developed torque and flux, if they are maintained at the corresponding reference values then the requisite torque is developed and flux is maintained. This would result in ω_s being equal to ω_{sREF} and the maximum wind energy being harvested. In a pulse-width modulator, v_{maREF}-v_{mbREF}-v_{mcREF} are compared with the carrier signal to generate the gate pulses G_{m1}-G_{m6} for the machine end converter.

5.5.5 Control of grid end converter

VOC is employed for the control of a grid end inverter. The theory of VOC remains the same as discussed earlier for control of the grid-tied photovoltaic system. Figure 5.19 shows the VOC-based control of a grid-end converter. A PLL algorithm is used to estimate the phase of supply voltage, $\omega_1 t$, based on v_{Pa}-v_{Pb}-v_{Pc}. This information is required for carrying out abc-to-dq0 and dq0-to-abc transformations. i_{wa}-i_{wb}-i_{wc} undergo an abc-to-dq0 transformation for determining i_{wd}-i_{wq}, which represent the active and reactive power being fed to the grid by the wind energy conversion system. The outer dc-link voltage controller processes the deviation in v_{wdc} from its reference value, v_{wdcREF}. The deviation, e_{vdc}, is further considered as an input for the dc-link voltage PI controller. This controller computes the reference value of the d-axis current to be supplied to the grid, i_{wdREF}, so that v_{wdc} is maintained at v_{wdcREF}. Here, v_{wdcREF} is determined based on the peak value of the grid voltage. In the absence of the reactive power injection, i_{wqREF} is held at zero. The inner d-q axes current control loop involves respective comparison of i_{wd}-i_{wq} with their reference values for the computation of d-q axes current errors, e_{id}-e_{iq}. e_{id}-e_{iq} are processed by separate current PI controllers for computing the reference d-q axes voltage, v_{mdREF}-v_{mqREF}. v_{mdREF}-v_{mqREF} are the reference d-q axes voltages to be maintained across the ac terminals of a grid end converter. v_{mdREF}-v_{mqREF} are forced to undergo adq0-to-abc transformation for obtaining v_{maREF}-v_{mbREF}-v_{mcREF}. Thus, obtained reference voltages in an a-b-c reference are further processed to generate the gate signals G_{w1}-G_{w6} for the grid end converter. The operation of a grid end converter with gate pulses G_{w1}-G_{w6} determined by VOC would result in IEE-1547 compliant operation with the injected current being sinusoidal with THD within limits and active power being injected into the grid, while the v_{dc} is maintained at the reference value.

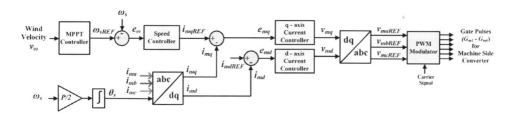

Figure 5.18 Block diagram representation of FOC-based control of a machine end converter.

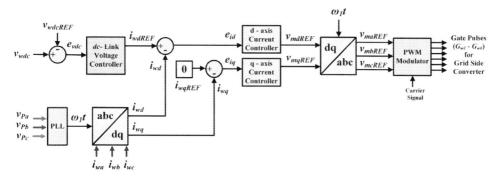

Figure 5.19 Block diagram representation of VOC-based control of a grid end converter.

5.6 RENEWABLE ENERGY FOR BUILDINGS

Buildings consume electricity for lighting, heating, air-conditioning, elevators, refrigeration, entertainment, water pumping, etc. In order to reduce the carbon emissions from buildings and to reduce grid reliance, renewable energy-based electric power generation can be incorporated into the buildings. Solar panels and wind turbines can be mounted on the rooftops and in gardens within building compounds. Solar trees can also be constructed. Vertical axis wind turbines are preferred in buildings as they can be placed on ground and are less noisy. The geographical location of the building and availability of space will play a huge role in selection of the renewable energy source. The algorithms and power structures discussed in the earlier sections are implemented to harvest wind and solar energy and feed it to the building loads and grid. There is a scenario where the power generated by the renewable energy source is in excess of the requirement or the generated power is less than the load requirement. In such cases, energy storage elements can be incorporated to develop a hybrid energy system. The excess power generated from the renewable energy source can be used to charge a battery, and this stored energy can be used to power the load at peak load periods or during nighttime when solar energy is not available. Figure 5.20 shows the hybrid energy system comprising of a single-stage grid-tied PV system with battery energy storage. The battery is interfaced at the dc-link by means of a bidirectional dc–dc converter. Bidirectional dc–dc converters facilitate controlled charging as well as controlled discharging of a battery pack. In the absence of a battery pack, the excess generated power can be fed back to the grid. The inclusion of net metering can result in revenue generation for building residents.

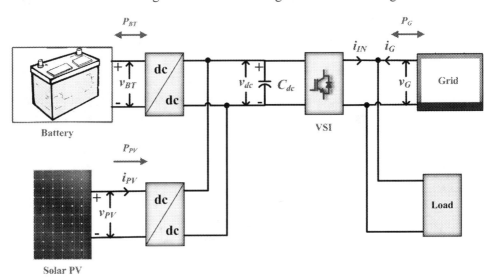

Figure 5.20 Hybrid energy system comprising a solar PV system and battery pack.

REFERENCES

Bakirtas, T., & Akpolat, A. G. (2018). The relationship between energy consumption, urbanization, and economic growth in new emerging-market countries, *Energy*. Vol. 147, 15 March 2018, pp. 110–121, doi.org/10.1016/j.energy.2018.01.011.

Burke, P. J., & Csereklyei, Z. (2016). Understanding the energy-GDP elasticity: A sectoral approach. *Centre for Applied Macroeconomic Analysis*. p. 38.

Gielen, D., Boshell, F., Saygin, D., Bazilian, M. D., Wagner, N., & Gorini, R. (2019). The role of renewable energy in the global energy transformation, *Energy Strateg. Rev.* Vol. 24, April 2019, pp. 38–50, doi.org/10.1016/j.esr.2019.01.006.

Goel, M. (2016). Solar rooftop in India: Policies, challenges and outlook. *Green Energy Environ.* Vol. 1, Iss. 2, pp. 129–137. doi.org/10.1016/j.gee.2016.08.003.

Harish V. S. K. V., & Sant A. V. (2020). Grid integration of wind energy conversion systems. In: Pathak P., Srivastava R. R. (eds) *Alternative Energy Resources. The Handbook of Environmental Chemistry*. Vol 99. Springer, Cham. doi.org/10.1007/698_2020_610.

Hayter, S. J., & Kandt, A. (2011). Renewable energy applications for existing buildings. *48th AiCARR Int. Conf.* September. pp. 1–15. Available: www.nrel.gov/docs/fy11osti/52172.pdf.

Horowitz, K., Peterson, Z., Coddington, M., Ding, F.,Sigrin, B., Saleem, D., Baldwin, S. E., Lydic, B., Stanfield, S. C., Enbar, N., Coley, S., Sundararajan, A., & Schroeder, C. (2019). An Overview of Distributed Energy Resource (DER) Interconnection: Current Practices and Emerging Solutions. Golden, CO. National Renewable Energy Laboratory. NREL/TP-6A20–72102. www.nrel.gov/docs/fy19osti/72102.pdf.

How a Wind Turbine Works – Text Version. Accessed on: May. 5, 2021. Available: www.energy.gov/eere/wind/inside-wind-turbine.

IEEE. (2003). 1547-2003 - IEEE standard for interconnecting distributed resources with electric power systems. *IEEE Std 1547–2003*. pp. 1–28, doi.org/10.1109/IEEESTD.2003.94285.

IEEE. (2018). 1547-2018 - IEEE standard for interconnection and interoperability of distributed energy resources with associated electric power systems interfaces. *IEEE Std 1547–2018 (Revision of IEEE Std 1547–2003)*. 6 April 2018. pp. 1–138. doi.org/10.1109/IEEESTD.2018.8332112.

IRENA. (2019). Global energy transformation: A roadmap to 2050 (2019 edition). *International Renewable Energy Agency*. Abu Dhabi.

IRENA. (2021). Renewable capacity statistics 2021. *International Renewable Energy Agency*. Abu Dhabi.

Kamran, M., Mudassar, M., Fazal, M. R., Asghar, M. U., Bilal, M., &Asghar, R. (2020). Implementation of improved Perturb & Observe MPPT technique with confined search space for standalone photovoltaic system. *J. King Saud Univ. - Eng. Sci.* Vol. 32, Iss. 7, pp. 432–441. doi.org/10.1016/j.jksues.2018.04.006.

Karad, S., & Thakur, R. (2021). Recent trends of control strategies for doubly fed induction generator based wind turbine systems: A comparative review. *Arch. Comput. Methods Eng.* Vol. 28, no. 1, pp. 15–29. doi.org/10.1007/s11831-019-09367-3.

Mao, M., Cui, L., Zhang, Q., Guo, K., Zhou, L., & Huang, H. (2020). Classification and summarization of solar photovoltaic MPPT techniques: A review based on traditional and intelligent control strategies. *Energy Reports*. Vol. 6, pp. 1312–1327. doi.org/10.1016/j.egyr.2020.05.013.

Masters, G. M. (2004). Wind power systems. *Renewable and Efficient Electric Power Systems*. John Wiley. pp. 307–383.

Masters, G. M. (2004). The Solar Resource. *Renewable and Efficient Electric Power Systems*. John Wiley. pp. 385–443.

Masters, G. M. (2004). Photovoltaic Systems. *Renewable and Efficient Electric Power Systems*. John Wiley. pp. 505–604.

Masters, G. M. (2004). Photovoltaic Materials and Electrical Characteristics. *Renewable and Efficient Electric Power Systems*. John Wiley. pp.445–504.

Mastromauro, R. A., Liserre, M., & Dell'Aquila, A. (2008). Study of the effects of inductor nonlinear behavior on the performance of current controllers for single-phase PV grid converters. *IEEE Trans. Ind. Electron.* Vol. 55, no. 5, pp. 2043–2052. doi.org/10.1109/TIE.2008.917117.

Nadimi, R., &Tokimatsu, K.(2018). Modeling of quality of life in terms of energy and electricity consumption. *Appl. Energy.*Vol. 212, 15 February 2018, pp. 1282–1294.doi.org/10.1016/j.apenergy.2018.01.006.

Njiri, J. G., & Söffker, D. (2016). State-of-the-art in wind turbine control: Trends and challenges. *Renew. Sustain. Energy Rev.* Vol. 60, pp. 377–393. doi.org/10.1016/j.rser.2016.01.110.

Nkuissi, H., Konan, F., Hartiti, B., & Ndjaka, J. (2020). Toxic materials used in thin film photovoltaics and their impacts on environment. *Reliability and Ecological Aspects of Photovoltaic Modules*, A. Gok, IntechOpen, doi.org/10.5772/intechopen.88326.

Qin, S., Wang, M., Chen, T., & Yao, X. (2011). Comparative analysis of incremental conductance and perturb-and-observation methods to implement MPPT in photovoltaic system. *2011 Int. Conf. Electr. Control Eng. ICECE 2011 - Proc.* pp. 5792–5795. doi.org/10.1109/ICECENG.2011.6057704.

Sant, A. V., Khadkikar, V., Xiao, W., Zeineldin, H., & Al-Hinai, A. (2013). Adaptive control of grid connected photovoltaic inverter for maximum VA utilization. *IECON Proc. Industrial Electron. Conf.* November 2013, pp.388–393, doi.org/10.1109/IECON.2013.6699167.

Sources of Greenhouse Gas Emissions. Accessed on: May. 5, 2021. Available: www.epa.gov/ghgemissions/sources-greenhouse-gas-emissions.

Tabassum, Z., & Shastry, C. (2022). Renewable energy Sector in Gujarat, India. *J Univ. Shanghai Sci. Technol.* Vol. 23, no. 6, pp.1128–1140.

UN. (2019). 2019 Global Status Report for Buildings and Construction Sector. The United Nations.

Chapter 6

Electric vehicle technology

Arpit J. Patel, Chaitali Mehta, Ojaswini A. Sharma, Amit V. Sant, and V.S.K.V. Harish

CONTENTS

6.1 NEED FOR ELECTRIC VEHICLES

Incessant carbon emissions have led to global warming and adverse climatic changes. This has resulted in increased awareness and interest in clean technologies. There has been an increasing rise in global greenhouse gas emissions, and a considerable contribution comes from the gaseous emissions of internal combustion engine-based vehicles. With the ideation of sustainable development, increased emphasis is being placed on the identification and development of cleaner and greener sources of energy. This has led to the search for green alternatives in all fossil fuel-based energy extensive domains. The automobile, since early civilization, has been an integral part of human development. Today, automobile manufacturers

DOI: 10.1201/9781003211587-6

across the globe are looking for cleaner alternatives to fuel the new generation of upcoming automobiles. E-mobility has therefore come to the forefront as a result and is driving transformation in the urban transportation sector.

A decade ago, the automobile sector was heavily based on internal combustion engine-based automobiles that run on petrol or diesel. The resulting emissions that are by-products of the exothermic combustion of these fuels produce significant carbon dioxide and carbon monoxide. Subsequently, compressed natural gas (CNG) was found to be a better alternative with lower carbon emissions and was thus introduced as a fuel for automobiles. This was a better alternative as the by-products of the combustion of CNG were water and carbon dioxide, thus eliminating other toxic carbon compounds from the emissions. With developments in semiconductor technology, material science, electrical machinery and energy storage, electric vehicles (EVs) are envisaged to grow leap and bounds in the coming decades.

6.2 ELECTRIC VEHICLE TECHNOLOGY

The early introduction of electrical components in automobile transmission was done with the introduction of electric motors and batteries along with mechanical components. Such power transmission was termed a hybrid power train. Fuel was burnt in the internal combustion engine, which was used to drive the vehicle using power transmission, and a regenerative braking power-recovery mechanism was added that charged the battery by extracting the power generally lost during braking. This power was then stored in the battery in the form of electrical power to be used for the auxiliary functions of the vehicle. When the stored electrical power was used for adding extra range to the vehicles, these vehicles were known as range-extended vehicles. While the hybrid power train provided the benefit of utilizing mechanical as well as electrical energy to drive the vehicle, the overall transmission efficiency of the system was very low and resulted in high energy losses. To eradicate these losses, a fully electric power train was designed. The electric power train offers a much better transmission efficiency of about 70%, as compared to the 30–40% of internal combustion engine-based vehicles. A new category of plug-in hybrid electric vehicles was also developed that provided the freedom of the user to charge the vehicle battery and use it for driving purposes or drive on traditional fuel combustion. Nowadays, battery electric vehicles have gained major focus because of their zero-carbon emission policy and higher transmission efficiency amongst the discussed categories.

Figure 6.1 shows the key components of an electric power train that propels a battery electric vehicle. The main components of an electric vehicle powertrain are:

a)   A battery pack
b)   A battery management system (BMS)
c)   Power modulators
d)   An electric motor
e)   A sensing unit and protection system

A detailed discussion on each of these components is provided below.

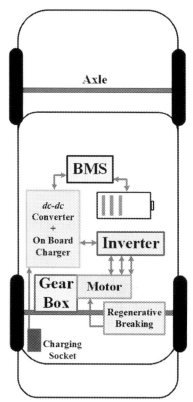

Rear Wheel Drive Battery Electric Vehicle

Figure 6.1 Electric propulsion system.

- **Battery pack:** The battery is the powerhouse of an electric vehicle. All the energy used to perform the different essential and auxiliary functions in a battery electric vehicle is sourced from the battery. An ideal battery for an electric vehicle must have high energy density, be compact, and be light in weight. It should also be able to function satisfactorily over a wide range of temperatures. Over time, various battery packs with different cell chemistries have been tried and tested, of which Li-ion cells show the most optimum performance for the required parameters. Today, most automobile manufacturers depend on Li-ion battery packs for their battery requirements. The battery pack employed in a battery electric vehicle is a series–parallel connection of cells for obtaining the requisite terminal voltages and ampere-hour capacity.
- **Battery management system:** The BMS is responsible for continuously monitoring the battery of the vehicle and managing battery health. It acts as a safety system for the protection of the battery and cuts off the battery supply if any hazardous situation

is detected. Since high power rating motors are used for propulsion in battery electric vehicles, high currents are drawn from the battery. The sudden drawing of such high currents may put additional stress on the battery, thus degrading its health in the long term. Therefore, a BMS is used to monitor the current being supplied to and from the battery in different scenarios and keeps a check on the current and voltage, and temperature limitations. Cell balancing can also be implemented with the help of BMS.

- *Power modulators:* Power modulators are essentially a group of different power converters that help in modulating or converting the electrical energy stored in the battery into the appropriate form of power required by the electric motor used in the vehicle. The power converters used in an electric propulsion system include the cascade connection of dc–dc converters and inverters for the appropriate power conversion process. Generally, electrical supply from the battery is provided to the dc–dc converters, which help in converting the available power to appropriate voltage levels. As ac motors are usually employed as propulsion motors, ac supply is needed. Further, to control the speed of the rotor shaft while avoiding magnetic saturation in the electric motor, it is essential to control the frequency and voltage of the stator supply. The three-phase variable voltage variable frequency supply is made available at the stator terminals through the dc–ac conversion facilitated by the voltage source inverters. dc–dc converters can play the additional role of on-board chargers for the charging of electric vehicles. Since the energy flow can be from the battery to the motor or from the external power supply to the vehicle battery while charging, dc–dc converters must support a bidirectional power conversion and supply. Hence, bidirectional dc–dc converters are used. Amongst the various topologies of bidirectional converters, the dual active bridge (DAB) converter has gained prominence due to its efficient power conversion control and isolated characteristics. The DAB consists of two full bridges, a high-frequency transformer, an energy transfer inductor, and dc-link capacitors. Control over the voltage level in a DAB converter is done by controlling the phase shift between the two bridges of the converter. The circuit diagram of a DAB is shown in Figure 6.2. The transformer shown in the diagram is a high-frequency ferrite core transformer, which helps in providing better magnetic coupling and electrical isolation along with compactness in size and minimum weight addition. The opposite pair of switches of a single bridge (S_{1P}, S_{3P} and S_{2P}, S_{4P}) are activated by providing gate pulses inversely to each pair (i.e., only one pair of a bridge is switched ON at a particular time interval to avoid a dead short-circuit condition in the circuit). Simultaneously, gate pulses are also provided to the switches of the second bridge switch pairs too (S_{1S}, S_{3S} or S_{2S}, S_{4S}) as per the switching logic. The switching table for the working of a DAB is shown in Table 6.1. The converters used in the power modulator employ power electronic switches such as MOSFETs and IGBTs as part of the power converters that supplement in providing a safe passage for the high electrical power being transferred for the electric propulsion.

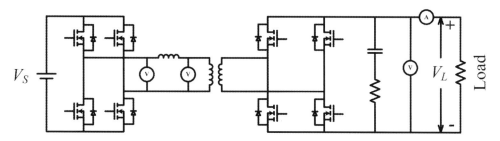

Figure 6.2 Dual active bridge dc-dc converter.

Table 6.1 Switching table of a dual active bridge converter

Interval	Switches of bridge 1 that are ON	Switches of bridge 2 that are ON	Voltage across inductor
I	S_{1P}, S_{4P}	S_{2S}, S_{3S}	$V_{dc1} + V_{dc2}$
II	S_{1P}, S_{4P}	S_{1S}, S_{4S}	$V_{dc1} - V_{dc2}$
III	S_{2P}, S_{3P}	S_{1S}, S_{4S}	$-V_{dc1} - V_{dc2}$
IV	S_{2P}, S_{3P}	S_{2S}, S_{3S}	$-V_{dc1} + V_{dc2}$

• **Electric motor:** Electric vehicles are propelled by electric motors. The power is modulated by power modulators using different converters and is made available at the stator terminals. The electric motor then utilizes this power to drive the vehicle. Earlier, dc motors were used for electric traction, but with advancements in machine design and the development of energy-efficient advanced electrical machines, they are largely employed as propulsion motors. Permanent magnet synchronous motors (PMSM), permanent magnet brushless dc (PMBLDC) motors, and switched as well as synchronous reluctance motors are popular as propulsion motors. These advanced motors require a power converter to control the stator winding excitation as per the rotor position for their proper working and speed control. Nowadays, most vehicle manufacturers prefer PMSM in four-wheel electric vehicles. However, some manufacturers also use efficient induction motors for this purpose. On the other hand, PMSM and PMBLDC motors are preferred in two-wheel and three-wheel electric vehicles.

• **Sensing unit and protection system:** Current sensors, voltage sensors, speed and position sensors, temperature sensors, proximity sensors, etc., are employed in electric vehicles for the control action required to ensure the desired motion of the vehicle and safety. Passenger and driver safety is a prime concern in any automobile, and consequently, many protections units are added in the form of electronically controlled units to continuously sense and monitor important parameters in the vehicle. These parameters include the currents being drawn for ongoing essential and auxiliary processes in the electric vehicle, temperatures at critical points, voltage levels, steering directions, braking force, etc. The electronic control unit continuously monitors these parameters and, on detection of an unsuitable situation, takes a series of control actions to bring the system parameters back to their normal ranges, thus increasing system stability and safety.

6.3 CHARGING INFRASTRUCTURE FOR ELECTRIC VEHICLES

Automobiles today are becoming "smart" with the increased inclusion of smart sensing, information and telecommunication technologies, and internet connectivity additions. Numerous features such as navigation and infotainment systems for consumer convenience and a luxurious experience are being added, along with advanced driver assistance systems (ADAS), vehicle-to-vehicle (V2V) and vehicle-to-infrastructure (V2I) communication systems for better navigation and tracking. Today, millions of vehicles are on-road in well-developed countries. Converting all these vehicles to electric vehicles will immensely help reduce a considerable amount of vehicular carbon emissions. While saving on carbon emissions, it will also help nations reduce their fossil fuel consumption and save on foreign exchange. Therefore, the adoption of electric vehicle technology will be a liberating process for any developing country today. But as with any new technology, there are challenges impeding the fast adaptation of electric vehicle technology. The range of an electric vehicle, along with the lack of electric charging station infrastructure, is a major cause of anxiety amongst electric vehicle customers. Countries today are working on several schemes and incentives for overcoming these technological challenges.

Charging infrastructure along with different charging methodologies are being developed and tested with the available power systems. The impact of charging electric vehicles, their impact on the grid, and scheduling electric vehicle charging are serious concerns that are being researched in academic and industrial circles. As opposed to direct fuel filling in conventional internal combustion engine-based vehicles, the electric charging of electric vehicles requires more time because of the limitations of battery parameters and degradation in battery health. Therefore, different charging methods and technologies are being tested to find an optimal solution that can be efficiently and economically implemented on a large scale for mass electric vehicle charging. Different charging methods and charger topologies are discussed in the following sections with their advantages and limitations.

6.4 CHARGING TECHNIQUES FOR ELECTRIC VEHICLES

Battery electric vehicles are powered by rechargeable battery. As the vehicles drive, battery power is consumed to generate tractive effort. The battery discharges and requires frequent charging. This, in turn, requires an investment in developing a robust and easily accessible charging network. One of the major hurdles in the widespread acceptance of electric vehicles is the range anxiety that can be reduced with the development of charging infrastructure across the breadth of a country. Along with a well-established charging infrastructure network, the type of charger selected also plays a key role. For a battery electric vehicle, the charging time is quite high: a two-wheel vehicle takes up to three hours to fully charge, whereas a four-wheel vehicle has a charging time of anywhere from 8 to 12 hours. This provision is available with slow ac chargers that can be used at home. To avoid congestion and reduce waiting time at public charging stations, a charger that can charge a vehicle in a much shorter time needs to be employed. A dc fast charger can provide this alternative. dc chargers charge the vehicle up to 80% in 30 minutes and can fully charge a four-wheel vehicle in less than two hours.

Depending on the physical location of the components of the charger utilized to charge the electric vehicle battery from the grid, chargers can be categorized as on-board or off-board. On-board chargers are located within the vehicle, and the available space within the vehicle determines the size and power rating of the charger. Off-board chargers are located outside the vehicle. More flexibility is available in off-board chargers in terms of deliverable power. Both on-board and off-board chargers must contain control circuits and need to communicate with the vehicle battery to ensure optimum charging of the battery. In this way, damage sustained by the battery due to overcharging can be completely avoided. On-board charging uses a slow ac charger, and off-board charging uses a dc fast charger. An inductive charger uses a combination of on-board and off-board chargers. Figure 6.3 depicts the concept of on-board and off-board chargers. The ac chargers are connected to the on-board charger. The on-board charger supplies power to the battery via a BMS. The task of the BMS is to protect the battery from overcharging.

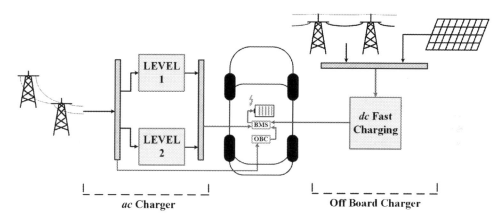

Figure 6.3 Onboard and off-board charger.

6.4.1 Conductive charging

Conduction is the phenomenon of energy transfer from one body to another through direct contact. A conductive charger can be further classified as an ac or a dc charger, discussed below.

6.4.2 ac charger

This charger allows an electric vehicle to be directly charged from ac plugs commonly available at home or office. ac charger operation includes the following:

a. The connection between the charger and the electric vehicle allows for communication between the two, and data related to the battery charging specifications and faults, if any, is carried out.
b. Rectification for ac–dc power conversion is followed by dc–dc power conversion to modulate voltage and current at its output for the controlled charging of the battery pack. In on-board charging, the power converters are located in the vehicle.
c. A BMS assists in battery charging by providing the necessary information regarding the state of the battery. The specific signals delivered by the BMS to the control unit are responsible for ensuring that no damage is sustained by the battery and its lifecycle is not compromised. The protection circuit isolates the battery if the need arises.

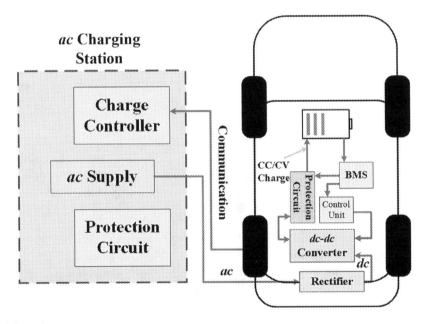

Figure 6.4 ac charger.

6.4.3 Level 1 charger

This is the slowest of all chargers in terms of charging time. It operates at the 120 V level and can be plugged into a standard 16 A socket commonly available at homes. Most electric vehicles are equipped with this charger. There is no additional cost of installation as this charger only requires a standard 16 A socket mounted on a wall. This charger has a nominal impact on the grid during its operation. However, the high charging time is a huge drawback associated with this charger.

6.4.4 Level 2 charger

This charger operates at 240 V for charging an electric vehicle battery. The charging time involved in this charger is less than a level 1 charger as the charging voltage is higher. Energy efficiency is also recorded to be higher. The limitations of level 2 chargers are a higher cost compared to a level 1 charger and a higher impact on the grid.

6.4.5 dc fast charger

A dc fast charger supersedes level 1 and level 2 chargers in terms of charging times. The output of a dc charger ranges from 50 kW to 350 kW. The high-power operation results in high cost and a complex converter design and control circuitry. This is why the dc fast charger is an off-board charger rather than an on-board charger. Using dc fast charging reduces charging time drastically and thereby addresses range anxiety for consumers. With a well-established charging station network equipped with dc fast chargers, the size of the battery can also be reduced, which in turn can reduce the pricing for electric vehicles. The limitation of using a dc fast charger is its high cost and high impact on the grid. Hence, it is most suited for public charging stations. Figure 6.5 shows the components involved in dc fast charging. dc fast charging involves a rectifier present in the charging equipment that converts the ac supply delivered by the grid into dc supply, which is further modulated by the dc–dc power converter for controlled charging of the battery pack. Provisions are included to protect the battery by halting battery charging in case of any abnormality in the battery or charger operation. The BMS monitors battery operation during the charging process and communicates with the control system for effective charging and protection of the battery.

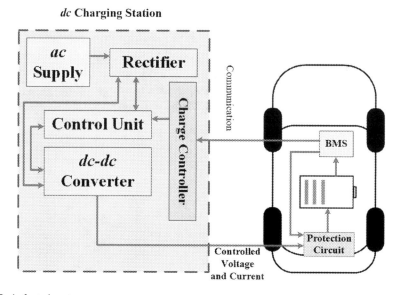

Figure 6.5 dc fast charging.

6.4.6 Inductive charging

Inductive charging is a form of contactless charging that involves electromagnetic induction between two electrically separated coils. The primary coil located on the road or parking lot is fed from a high-frequency ac supply. High-frequency ac voltage is induced in the secondary coil located at the bottom of the vehicle. The rectifier and dc–dc converter unit located in the electric vehicle modulate the power transferred to the secondary coil to ensure controlled charging of the battery pack. The concept of inductive charging is shown in Figure 6.6. The high-frequency ac supply is obtained by rectification of supply voltages followed by dc–ac conversion. In this inverter operation, the frequency of the output ac supply can be controlled. Moreover, there is no barrier from the end of the inverter in terms of limiting the frequency. This process is carried out off-board.

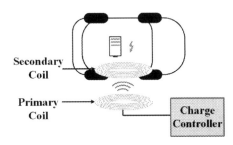

Figure 6.6 Concept of inductive charging.

There are two types of inductive charging, static and dynamic charging, which are discussed below.

6.4.7 Static charging

During charging, both vehicle and charger are stationary. Static charging can be done at home or an office parking lot just by upgrading the area. The hassle of wires can be removed. This type of system can be buried or flush-mounted, thereby not affecting the façade of the city and being safe due to its inaccessibility. The major limitations of this charger are the high investment cost and higher losses when compared to conductive charging.

6.4.8 Dynamic charging

Another way to charge the vehicle wirelessly is called dynamic charging. The secondary coil is still connected to the vehicle, as in static charging. The multiple primary coils connected to the power grid are buried along the road. In this way, the car can be charged even while it is in motion. The primary coil transfers the power to the secondary coil located inside the vehicle through electromagnetic induction. The induced electromotive force (emf) is converted to dc and utilized to charge the battery. The advantage of using dynamic charging is low stand-in charging time, low battery depth of discharge, and smaller battery size. The limitations are higher cost, interference from external elements, misalignment on the road, and applicability to different electric vehicles.

6.4.9 Battery swapping

The time required by ac or dc chargers to charge a battery pack is higher compared to the fuel-refilling time associated with internal combustion engine-based vehicles. The same is the case with inductive charging. Charging times are one of the key concerns for the wider acceptance of electric vehicles; immediate charging is not possible. However, charging stations can be renamed battery swapping stations. At battery swapping stations, the electric vehicle user can have the charge-depleted battery replaced with a fully charged battery. Battery swapping can be done in a manual or automated fashion. Battery charging can take place at the charging station without any constraints over the charging time. The concept of battery swapping is shown in Figure 6.7. Hence, slow charging can be carried out to extend the life of the battery. Additionally, a solar photovoltaic system can also be incorporated to charge the electric vehicle to further reduce reliance on the grid. Battery swapping is the fastest way to get a discharged battery replaced by a fully charged battery. It has generated a lot of interest, and work is in progress on formulating different business models for battery swapping. From a consumer perspective, the price of the electric vehicle would be reduced as the battery does not need to be procured but leased out at the swapping station. Lack of availability of common battery standards for different vehicles and a successful business model are drawbacks that need to be addressed.

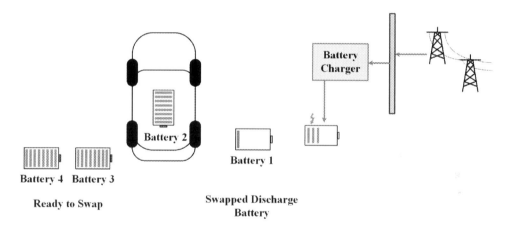

Figure 6.7 Battery swapping.

6.5 VEHICLE TO GRID (V2G)

Vehicle to grid (V2G) technology has been developed to utilize the electrical energy stored in an idle electric vehicle battery or any other energy storage device. It can assist with increased energy demands during peak load hours, maintain grid stability, avoid the need for peak power plants, etc. Further, it can help increase the indirect penetration of renewable energy sources when used for charging electric vehicle batteries. V2G can be defined as the transfer of power or energy from an electric vehicle battery to the grid using suitable power electronic converters for the purpose of managing load demand. Energy or power transfer using the V2G technique can be divided into four sections, as shown in Figure 6.8.

(i)   Extracting power from an electric vehicle battery
(ii) Power electronic converters
(iii) Control and measuring units
(iv) Transformers for injecting power into the grid

While extracting power from an electric vehicle battery, bi-directional dc–dc converters are used to extract and inject power from or to the battery. As the battery voltage is low compared to the output dc bus voltage, the dc–dc converter needs to boost the voltage levels. To convert the required power from dc to ac, inverters are employed. The ac terminals of the inverter are connected to the grid through passive filters that prevent the harmonic components from propagating into the grid. Furthermore, power transformers are used to interface the inverter with the grid so as to provide necessary isolation as well as step up the voltage as per the desired levels. The power converters require an activation signal in the form of gate pulses for the desired operation. Moreover, power converters are involved in controlling the flow of power, grid synchronization, and regulating the dc link voltage. The control algorithm for power converters is employed through a control unit, which necessitates the instantaneous values of different quantities for closed-loop control. A measurement unit involves different sensors for facilitating the instantaneous measurements of the required quantities.

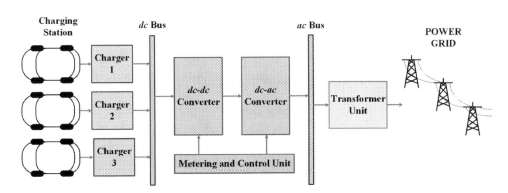

Figure 6.8 Block diagram of V2G.

As shown in Figure 6.8, electric vehicles are connected to charging stations, followed by power electronic converters and transformers for interfacing the dc–ac converter (i.e., inverter) with the grid. The measuring unit provides measurement signals to the control unit, which implements the control algorithm for the generation of the gating pulse for the desired action. With these connections, electric vehicles are charged during low-load demand, and as soon as the load demand increases, the battery of the electric vehicle acts as a distributed energy source for the grid and supplies power to the grid for managing load demand. When the energy production is high, the price of energy would be low. On the other hand, when the demand is high, the energy price would be high. By charging electric vehicle batteries during this low load demand and supplying power back to the grid during high load demand, electric vehicle owners can generate revenue and help the utility. Also, due to the intermittent nature of renewable energy and increased renewable energy generation, energy storage elements are required. Rather than incurring the cost of a centralized battery, electric vehicle batteries can be deployed. With all the benefits mentioned above, some challenges for implementing V2G technology include the requirement of a dedicated communication system, high capital cost, fast battery degradation, complex hardware, and less social awareness.

Electric vehicles owners, aggregator/grid operators, and utility grid operators are the three key players that enable the V2G concept. In the V2G system, the electric vehicle owner is fundamental since they own the vehicle and put it up for the V2G interface. The utility operator is responsible for balancing the transfer of information and electricity between the grid and the vehicle. The major function of the electrical grid operator is to transmit large-scale electricity generation to demand areas through high-voltage transmission lines, and to balance generation and load generally over large regional areas. The secondary actors for the V2G concept with electric vehicles are the electric vehicle industry, government, policymakers, and electricity producers.

6.6 IMPACT OF V2G ON THE GRID

Electric vehicle technology is a potential solution to the issue of carbon emissions caused by internal combustion engine-driven automobiles. Combining this with renewable energy-based generation appears to be a promising strategy for reducing greenhouse gas emissions in the energy and transportation sectors. The growing popularity of electric vehicles can certainly result in a significant increase in charging demand on the electric grid. Furthermore, the timing and duration of electric vehicle-induced loads is highly uncertain. These would bring supply-demand management problems in the electric grid, posing several potential risks to its stable and economic operation. Moreover, the high penetration of renewable energy systems into the grid combined with its intermittent nature poses vital challenges to the grid, including supply-demand mismatch and voltage and frequency instability.

Electric vehicle charging load can result in a mismatch between active power supply and demand, causing frequency variations. If the charging load is not planned and scheduled beforehand, these problems can become even more severe. Synchronous generators are equipped with load frequency control. The generated active power is regulated to maintain frequency stability. Traditional sources have constraints, such as a ramp rate limit for different types of power generation plants. Large generators are utilized to meet surplus active power demand and for frequency regulation. In particular, a spinning reserve of generators is usually kept on standby to handle any unexpected surges in demand. The problem with generators is that they require a certain amount of time to start and may not be able to deliver the load immediately if started cold. The V2G concept enables electric vehicles to be employed to perform additional services with appropriate supervision and control. Electric vehicles can be used for frequency regulation services by providing active power to the grid with the concept of V2G. Upon deviation in the grid frequency, V2G can absorb or supply energy to assist in frequency regulation. Furthermore, the V2G concept enables electric vehicles to assist in balancing grid supply and demand. For the peak load period, with V2G, the battery can supply the grid. Alternately, during low load periods, the battery can be charged. The characteristics that make electric vehicles suitable for participating in grid frequency control are quick response, ease of operation, efficient energy conversion, etc.

With the widespread and increasing usage of electric vehicles, extensive battery charging load may impose challenges such as voltage fluctuation. Furthermore, the random charging of electric vehicles might overload transformers and increase energy losses in distribution network feeders. The planning of an optimal and coordinated charging scheme is crucial for the seamless functioning of the grid. Electric vehicles, on the other hand, can be used to provide voltage support in the distribution network by discharging their batteries into the grid during peak demand periods while ensuring that enough energy is stored until the next charging event. Moreover, the charging of electric vehicles with renewable energy resources and thereafter delivering power to the grid increases the penetration of renewable energy resources into the grid indirectly. In some cases, when the generation from the solar photo-voltaic system exceeds the local demand at the point of common coupling, issues such as reverse power flow and voltage rise may occur. In this situation, integrating storage devices with solar photovoltaic systems to store excess energy from a solar photovoltaic system by charging batteries could be an appealing alternative. The V2G concept can be extended to charging stations, battery energy storage systems, and battery swapping stations. It can be incorporated in buildings as well. In common parking lots, V2G units can be placed for residents. Similarly, in offices, V2G connections can be provided for employees to assist the grid and earn revenue.

6.7 BENEFITS OF V2G

The V2G concept offers potential benefits to electric vehicles users as well as the utility grid. The usage of electric vehicles for energy arbitrage is also possible. Electric vehicle owners can have savings on energy bills and generate revenue. This is possible if charging the battery during low-tariff periods and supplying power to the utility during high-tariff periods. The battery can also power the domestic loads for high-tariff periods. During low-load periods, the storage capacity of electric vehicles can be utilized to store excess power generated by renewable energy units. This stored energy can help power the utility during periods when heavy demand is experienced by the grid. Further, with V2G, electric vehicle batteries can support the grid for voltage as well as frequency regulation. Compensation of reactive power and load balancing by valley filling can also be affected with V2G. Aside from the technical advantages, the V2G concept, which allows electric vehicle owners to generate revenue, encourages more people to purchase electric vehicles. This has broader implications for carbon footprint reduction in transportation and, to some extent, energy generation considering the incorporation of renewable energy resources.

6.8 CHALLENGES FOR IMPLEMENTATION OF V2G

Due to pollution, greenhouse gas emissions, global warming, and climate change, there is a shift from fossil fuel-based technology to clean energy technology. Electric vehicles are ideally suited for this situation as they offer zero emissions while moving around from one place to another. Also, to limit the gaseous emissions from thermal power plants, grid integration of renewable energy sources is being deployed heavily. The batteries used in electric vehicles to power the generation of the required tractive effort for propulsion can also be used as a mobile energy bank. In situations of high demand for energy, it can transfer energy from battery to grid to mitigate peak demand. This concept is called vehicle-to-grid (V2G) energy transfer.

The major challenges in the development of V2G technology are financial challenges, uncertain behaviour of electric vehicles, reduction in battery life cycle, immature market for adoption of V2G technology, and social challenges.

1. **Financial challenges**: This might not be a big issue for developed countries, but it is a major issue for developing and underdeveloped countries. V2G technology involves electric vehicle owners, aggregators, and grid operators, and it is necessary to incorporate all three of these for the seamless running of processes. In order to do so, a well-established infrastructure is required with various facilities such as efficient telecommunication, metering input facilities, output power and plug-in connectors, and uncertainty in revenue generation.
2. **Uncertain behaviour of electric vehicles**: Uncertain behaviour in departure/arrival time to charging stations, driving patterns, battery sizes, and type and size of charger are the parameters with the greatest effect on grid stability and reliability.
3. **Reduction in battery life cycle**: With frequent charging and discharging, voltage and temperature can reduce the life of a battery. To motivate electric vehicle owners to actively participate in V2G, the cost of battery per kWh needs to be reduced, and the number of charging and discharging cycles needs to be increased.
4. **Immature market for adoption of V2G technology**: The share of electric vehicles is very low, around 1.3% globally, which limits electric vehicle owners and awareness of electric vehicles technology compared to fossil fuel-based vehicles. With a very low market share, very few original product manufacturers are available to provide V2G technology.

REFERENCES

Alam, M. J. E., Muttaqi, K. M., & Sutanto, D. (2016). Effective Utilization of Available PEV Battery Capacity for Mitigation of Solar PV Impact and Grid Support with Integrated V2G Functionality. *IEEE Trans. Smart Grid*. Vol. 7, No. 3, pp.1562–1571. doi.org/10.1109/TSG.2015.2487514.

Bibak B., & Tekiner-Moğulkoç, H. (2021). A comprehensive analysis of Vehicle to Grid (V2G) systems and scholarly literature on the application of such systems. *Renew. Energy Focus*. Vol. 36, pp.1–20. doi.org/10.1016/j.ref.2020.10.001.

Dharmakeerthi, C. H., Mithulananthan, N., & Saha, T. K. (2014). Impact of electric vehicle fast charging on power system voltage stability. *Int. J. Electr. Power Energy Syst*. Vol. 57, pp.241–249. doi.org/10.1016/j.ijepes.2013.12.005.

Ehsani, M., Gao, Y., Longo, S., & Ebrahimi, K. (2018). *Modern Electric, Hybrid Electric, and Fuel Cell Vehicles*. 3rd ed. CRC Press, Taylor and Francis Group.

Ghotge, R., Van Wijk, A., & Lukszo, Z. (2019). Challenges for the design of a Vehicle-to-Grid Living Lab. *Proc. 2019 IEEE PES Innov. Smart Grid Technol. Eur. ISGT-Europe 2019*. pp.1–5. doi.org/10.1109/ISGTEurope.2019.8905503.

Habib, S., Kamran, M., & Rashid, U. (2015). Impact analysis of vehicle-to-grid technology and charging strategies of electric vehicles on distribution networks: A review. *J. Power Sources*. Vol. 277, October, pp.205–214. doi.org/10.1016/j.jpowsour.2014.12.020.

Husain, I. (2003). *Electric and Hybrid Vehicles: Design Fundamentals*, CRC Press.

Li, M., & Lenzen, M. (2020). How many electric vehicles can the current Australian electricity grid support? *Int. J. Electr. Power Energy Syst*. Vol. 117, November 2020, p.105586. doi.org/10.1016/j.ijepes.2019.105586.

Malmgren, I. (2016). EVS29 International Battery, Hybrid and Fuel Cell Electric Vehicle Symposium Quantifying the Societal Benefits of Electric Vehicles./ *World Electr. Veh. J.* Vol. 8, pp.996–1007.

Mazumder, M., & Debbarma, S. (2021). EV Charging Stations with a Provision of V2G and Voltage Support in a Distribution Network. *IEEE Syst. J.* Vol. 15, No. 1, pp.662–671. doi.org/10.1109/JSYST.2020.3002769.

Nguyen, H. N. T., Zhang, C., & Zhang, J. (2016). Dynamic demand control of electric vehicles to support power grid with high penetration level of renewable energy. *IEEE Trans. Transp. Electrif.* Vol. 2, No. 1, pp.66–75. doi.org/10.1109/TTE.2016.2519821.

Un-Noor, F., Padmanaban, S., Mihet-Popa, L., Mollah, M. N., & Hossain, E. (2017). A comprehensive study of key electric vehicle (EV) components, technologies, challenges, impacts, and future direction of development. *Energies.* Vol. 10, No. 8, pp.1–82. doi.org/10.3390/en10081217.

Vadi, S., Bayindir, R., Colak, A. M., & Hossain, E. (2019). A review on communication standards and charging topologies of V2G and V2H operation strategies. *Energies.* Vol. 12, No. 19. doi.org/10.3390/en12193748.

Zhong, J., He, L., Li, C., Cao, Y., Wang, J., Fang, B., Zeng, L., & Xiao, G. (2014). Coordinated control for large-scale EV charging facilities and energy storage devices participating in frequency regulation. *Appl. Energy.* Vol. 123, pp.253–262. doi.org/10.1016/j.apenergy.2014.02.074.

Chapter 7

Techno-economic analysis of electric vehicles

Chaitali Mehta, Amit V. Sant, Arpit J. Patel, and V.S.K.V. Harish

CONTENTS

7.1 ELECTRIC VEHICLES

With electric vehicles offering operation with zero carbon emissions, more emphasis is being placed on increasing their acceptance to deal with environmental issues. The key economic considerations when buying an electric vehicle are its purchase cost, operating cost, and maintenance cost. The development of necessary infrastructure is critical to accelerating the adoption of electric mobility. Several government initiatives are helping to encourage the purchase of electric vehicles and the development of charging stations in this regard. Tremendous technological advancements in battery technology, propulsion systems, and charging infrastructure over the past decade have resulted in reduced electric car costs and, as a result, increased global adoption. The data for four-wheel electric vehicle sales in the past decade is plotted in Figure 7.1, as per IEA's *Global EV Data Explorer* (2021). In 2020, a 30% increase in battery electric vehicle car sales is seen compared to the previous year. Hybrid electric car sales also increased by 69.79%. A 40.85% increase in the sale of pure and hybrid electric vehicles was observed in 2020 as compared to that recorded in 2019. The countries with the highest four-wheel electric vehicle sales in 2020 are presented in Figure 7.2 as per IEA's *Global EV Data Explorer* (2021).

In India, the total number of electric vehicles sold in 2021 were 93,694, of which 73,254 were two-wheel, 18,841 were three-wheel, and 1,594 were four-wheel vehicles, as per FAME India Scheme Phase II, National Automotive Board (NAB), Ministry of Heavy Industries, Government of India. The increase in electric vehicle sales is observed in India as a result of incentives on buying electric vehicles by the central government and various state governments. The state-wise sale of electric vehicles is listed in Table 7.1.

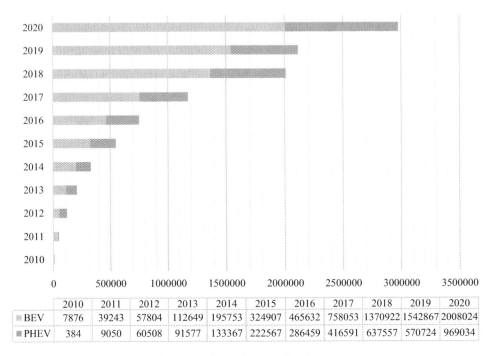

	2010	2011	2012	2013	2014	2015	2016	2017	2018	2019	2020
BEV	7876	39243	57804	112649	195753	324907	465632	758053	1370922	1542867	2008024
PHEV	384	9050	60508	91577	133367	222567	286459	416591	637557	570724	969034

Figure 7.1 Four-wheel electric vehicle car sales in the past decade.

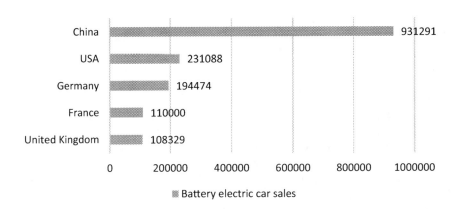

Figure 7.2 Four-wheel electric vehicle car sales in different countries.

Table 7.1 State-wise sale of electric vehicles in India

Andaman and Nicobar (UT)	2	Kerala	2,202
Andhra Pradesh	3,580	Lakshadweep (UT)	4
Assam	415	Madhya Pradesh	3,098
Bihar	2,771	Maharashtra	10,105
Chandigarh (UT)	52	Manipur	78
Chhattisgarh	2,146	Meghalaya	6
Dadra and Nagar Haveli (UT)	27	Orissa	1,806
Daman and Diu (UT)		Puducherry (UT)	160
Delhi	6,817	Punjab	880
Goa	260	Rajasthan	7,338
Gujarat	1,704	Tamil Nadu	14,678
Haryana	1,580	Telangana	3,195
Himachal Pradesh	250	Tripura	544
Jammu and Kashmir	487	Uttar Pradesh	6,343
Jharkhand	867	Uttarakhand	1,099
Karnataka	20,410	West Bengal	788

7.2 ELECTRIC VEHICLE TECHNOLOGY: MERITS AND CHALLENGES

Depending on the structure of the power train, electric vehicles are generally categorized as:

a. hybrid electric;
b. plug-in hybrid electric;
c. battery electric;
d. range extender; and
e. fuel cell electric vehicles.

From the point of view of carbon emissions, battery electric vehicles have the highest potential for reducing carbon emissions. A battery-operated electric vehicle has zero gaseous emissions when it is moving on the road. Hence, from the point of view of sustainability, reduced carbon emissions, and reducing global warming, the shift needs to be from internal combustion engine-based vehicles to hybrid vehicles and subsequently to battery electric vehicles. Carbon emissions are chiefly responsible for global warming and climate change. With increased awareness in this regard, significant efforts are being made at the global level for the development of technology that can contribute to sustainable development. The potential for reduced carbon emissions is one of the key factors for the interest in electric vehicle technology.

In battery electric vehicles, the propulsion motor is powered by a battery. The flow of electric power from the battery to the propulsion motor is modulated by power converters. With recent developments in power and signal electronics, solid-state drives, and electric motors, precise control over the developed torque and rotor speed can be easily achieved. The electric propulsion system provides instantaneous control over the developed torque and high responsiveness to the driver commands. Moreover, with only the electric motor as the moving part, the electric propulsion system is more efficient compared to internal combustion engine-based drive trains. The cost per kilometre incurred when driving electric vehicles is far less than that incurred when driving internal combustion-based vehicles. Furthermore, governmental policies have a critical role in raising public awareness towards carbon emissions and global warming, as well as the wider acceptance of electric vehicles. Subsidies, taxation policies, and preferential treatments for electric vehicles (freedom from toll taxes, permission to drive in bus lanes, etc.) contribute to the public preference for electric vehicles.

In spite of several measures, there are several issues that have played on the mind of the general public and prevent the widespread market acceptance of electric vehicles:

a. **Charging electric vehicle batteries**: Batteries supply electric power for propulsion. The battery discharges as the electric vehicle moves from one location to the other; hence, batteries require frequent charging. Slow-charging and fast-charging options are available for electric vehicle batteries. While the petrol or diesel tanks of electric vehicles can be filled within a few minutes, the charging of electric vehicle batteries takes a significant amount of time. Charging these batteries can take from an hour to as high as 12 hours or more. This is a big concern that plays on the mind of the public. The ease of availability of charging stations is another concern.

b. **Impact on the grid**: The simultaneous charging of electric vehicle batteries would present a huge challenge as it would increase the peak demand and reduce reserve margins. Further, battery charging can degrade the power quality of the grid, increase stability issues, and decrease system efficiency. The impacts of battery charging on indices and the economics of the power system need to be analyzed.

c. **Need for battery replacement**: Electric vehicle batteries are costly and form a significant part of the total cost. Usually, manufacturers give a warranty on electric vehicle batteries for five years. In some cases, this warranty may be for eight years. Generally, electric vehicle batteries need replacement after about five years. The life of a battery depends on its charge-discharge cycle. When the battery is charged it undergoes a charge cycle, and when it discharges it undergoes a discharge cycle. With every charge-discharge cycle, the health of the battery deteriorates. This implies that the charge that can be stored by a battery reduces. Consequently, the driving range for the full charge of the battery also reduces. Deviation in the specified norms for charging the battery can also adversely affect the health of the battery. The high cost of battery replacement and the period over which the battery needs to be changed is a key concern for consumers.

d. **Cost of electric vehicle**: The battery is one of the key components of the electric vehicle. As mentioned earlier, batteries are costly, and the battery pack forms one of the major components of the cost of an electric vehicle. Permanent magnet brushless motors are employed as propulsion motors in electric vehicles. These motors utilize Nd-Fe-B magnets. The increasing cost of these magnets is another cause of concern. Presently, electric vehicles are costlier as compared to internal combustion engine-based vehicles. With the mass production of electric vehicles and government subsidies, the cost of electric vehicles is bound to reduce and be comparable to internal combustion engine-based vehicles. Reduced costs would be a huge factor in the wider acceptance of electric vehicles.

7.3 TECHNO-ECONOMIC ANALYSIS OF ELECTRIC VEHICLES

Techno-economic analysis, also known as techno-economic assessment (TEA), evaluates a product, process, or service with a focus on economic aspects. It weighs the technological advancement against the financial costs. A mathematical model is developed first based on which parameters such as initial investment, running cost, and business model are analyzed. A TEA is typically used in the following examples:

a. **Evaluating the economic feasibility of a technology**: With a TEA, the developed technology can be analyzed and evaluated to determine whether it can affect economic gains for the given set of conditions.
b. **Guiding research and development**: Sensitivity analysis can help determine the impact of a variation in output variables with the change input variables. This analysis, when carried out with a TEA, can help in segregating the targeted research outcomes with the highest chances of yielding financial gains.
c. **Quantifying uncertainty and risk**: When dealing with financial aspects related to business, uncertainty in the model needs to be taken into account. Sensitivity analysis can be employed for determining economic uncertainty. The parameters that impact the model uncertainty the most can also be detected.

A TEA is used to assess the economic viability of any technological development being launched. Different factors related to economics and technology affect the market acceptance of electric vehicles and the further development of this technology. Figure 7.3 shows the schematic diagram of factors used for the TEA of an electric vehicle. Presently, the capital cost of an electric vehicle is higher than an internal combustion engine-based vehicle. The higher price is a result of various factors, such as lower manufacturing cost of internal combustion engine-based vehicles, an efficient assembly line for internal combustion engine manufacturing, higher battery price, etc. However, with advancements in battery technology, the capital cost of an electric vehicle is decreasing incessantly. It is expected to further reduce with large-scale mass production. The role of government agencies also plays a significant role in the early adoption of electric vehicles. These

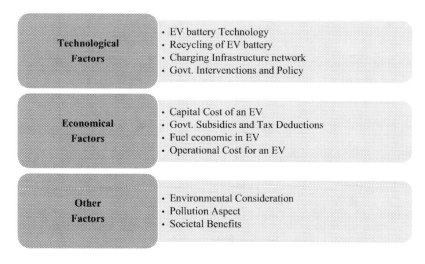

Figure 7.3 Factors to be considered for the TEA of an electric vehicle.

agencies can incentivize early adoption by providing economic benefits in terms of subsidies on manufacturing, vehicle purchase, charging infrastructure setup, tax deductions, etc. A new technology is said to be successful if the increased cost is matched by the technological advancement.

For any vehicle, the total cost of ownership (TCO) is considered when determining its economic burden on its owners. The TCO is calculated by considering all the expenses that can occur when we drive a vehicle. These expenses include capital cost, fuel charges, operation and maintenance cost, battery replacement cost, etc. By comparing the TCO of an electric vehicle with its counterpart internal combustion engine vehicle, it can be deduced that electric vehicle technology will be easily adopted by the general public in abundance only when the TCO of an electric vehicle is equivalent to the TCO of an internal combustion engine vehicle.

In terms of technological advancements, electric vehicle technology is farther ahead than the internal combustion engine vehicle. The efficiency of a propulsion motor is three to four times higher than the efficiency of the engine in an internal combustion engine-based vehicle. The maintenance cost of an electric vehicle is also minuscule when compared to an internal combustion engine vehicle. The ever-rising cost of petrol and diesel further strengthens the case for electric vehicle adoption. There have also been massive advancements in battery technology. The current choice for electric vehicle manufacturers is lithium-ion based batteries; these batteries have higher energy density, specific energy, safety, longer lifecycle, better performance, etc., when compared to lead-acid batteries. The cost of lithium-ion batteries is also dramatically reducing. Another technological issue in electric vehicle adoption is a lack of fast chargers for lead-acid batteries; however, with lithium-ion batteries, this issue is resolved as these batteries provide better C-rating charging and discharging capabilities.

Other significant factors that should be kept in mind while performing a TEA of electric vehicles are their environmental friendliness and emission-free operation. In India, air pollution is a serious concern. Several measures are being put into practice for reducing and controlling air pollution within cities. If the transportation sector shifts from internal combustion engine-based vehicles to electric vehicles, air pollution can be considerably reduced. The reduction in air pollution will also lead to better and healthier lifestyles for the people within the city.

7.4 COST COMPETITIVENESS OF ELECTRIC VEHICLES

The approximate cost comparison of a two-wheel electric vehicle and an internal combustion engine-based vehicle in India in terms of capital cost is presented in Table 7.2. Similarly, Tables 7.3 and 7.4 show the approximate cost comparison of a two-wheel electric vehicle and an internal combustion engine-based vehicle in India in terms of operation parameters and operation costs. The following assumptions are considered for comparison.

  a. The lifetime of both vehicles is taken as ten years.
  b. The lifetime of a lead-acid battery is three years (deep-cycle lead-acid battery).
  c. The lifetime of a Li-ion battery is six years.
  d. The battery is 50% of the cost of the electric vehicle.

Table 7.2 Approximate cost comparison of a two-wheel electric vehicle and an internal combustion engine-based vehicle in India in terms of capital cost

Capital cost component	Internal combustion engine-based two-wheel vehicle	Two-wheel electric vehicle
Base price	70,000 INR	75,000 INR
Goods and services tax	20,000 INR	3,750 INR
State goods and services tax	10,000 INR	Exempted
Central goods and services tax	10,000 INR	3,750 INR
Vehicle tax	1,250 INR	Exempted
Insurance	5,500 INR	5,500 INR
Other charges	1,450 INR	1,450 INR
Type of battery	–	Lithium-ion
Battery capacity (kWh)	–	1.875
Cost of battery	–	29,500 INR
Escalation rate of battery price (%)	–	10.00%

Table 7.3 Approximate cost comparison of a two-wheel electric vehicle and an internal combustion engine-based vehicle in India in terms of operation parameters

Operation parameters	Internal combustion engine-based two-wheel vehicle	Two-wheel electric vehicle
Annual distance travelled	10,000 km	10,000 km
Range	300 km	80 km
Mileage	60 km/litre	42.75 km/litre
Fuel cost	90.93 INR/litre	
Escalation of fuel cost	5.00%	
Electricity tariff		15.00 INR/Unit
Escalation of electricity tariff		5.00%

Table 7.4 Approximate cost comparison of a two-wheel electric vehicle and an internal combustion engine-based vehicle in India in terms of operation costs

Operation costs	Internal combustion engine-based two-wheel vehicle	Two-wheel electric vehicle
Cost of tyres	1,500 INR	2000 INR
Insurance premium	5,000 INR	5,000 INR
Engine oil type	Semi-synthetic oil	–
Engine oil cost	250 INR/litre	–
Oil filter	200 INR	–
Battery	1000 INR	–
FAME subsidy	–	17,000 INR
State government subsidy	–	9,360 INR
Electricity tariff with solar photovoltaic based generation	–	2.50

Based on the comparisons presented, the following conclusions are drawn:

a.   The levelized cost of a two-wheel electric vehicle is 1.56 INR/km
b.   The levelized cost of an internal combustion engine-based two-wheel vehicle is 3.42 INR/km
c.   The FAME subsidy is 17,000 INR
d.   The state government subsidy is 9,360 INR
e.   The total subsidy is 26,360 INR

The approximate total cost (including capital cost, operation and maintenance cost, fuel cost, and battery replacement cost) of an internal combustion engine-based two-wheel vehicle and two-wheel electric vehicle over a ten-year period is 342,000 INR and 156,000 INR, respectively. For each year over this ten-year period, it is assumed that the two-wheel vehicle covers a distance of 1000 km. Considering this duration and distance travelled per year, the cost incurred per kilometre of distance travelled is 3.42 INR and 1.56 INR for internal combustion engine-based two-wheel vehicles and two-wheel electric vehicles, respectively. This cost per kilometre includes the capital cost as well. Over a ten-year period, with 1,000 kilometres travelled each year, the two-wheel electric vehicle costs about half that of the internal combustion-engine based two-wheel vehicle. Thus, significant economic savings can be achieved with electric vehicles.

REFERENCES

IEA. (2021). *Global EV Data Explorer*, IEA, Paris https://www.iea.org/articles/global-ev-data-explorer

Kennedy, D., & Philbin, S. P. (2019). Techno-economic analysis of the adoption of electric vehicles. *Front. Eng. Manag.* Vol. 6, No. 4, pp. 538–550.

National Automotive Board (NAB) Ministry of Heavy Industries, Government of India. https://fame2.heavyindustry.gov.in/ (01/08/2021, 9:00 PM).

Swanson, R. M., Platon, A., Satrio, J. A., & Brown, R. C. (2010). Techno-economic analysis of biomass-to-liquids production based on gasification. *Fuel.* Vol. 89, No. SUPPL. 1, pp. S11–S19.

Yang, H., Wei, Z., & Chengzhi, L. (2009). Optimal design and techno-economic analysis of a hybrid solar-wind power generation system. *Appl. Energy.* Vol. 86, No. 2, pp. 163–169.

Zoulias, E. I., & Lymberopoulos, N. (2007). Techno-economic analysis of the integration of hydrogen energy technologies in renewable energy based stand-alone power systems. *Renewable Energy.* 32(4): 680–696.

Index

Printed in the United States
by Baker & Taylor Publisher Services